Seed Fungi:

Identification Character

2020

DR. A .K. KUSHWAHA

M.Sc., B.Ed., M.A., Ph.D.

Department of Botany,

Sarla Dwivedi P.G College
Akbarpur Kanpur (D) U.P India

ISBN: **9798640341447**

CONTENTS

.1. *Alternaria alternata* (Fries.) Keissler (Syn. *A. tenuis* Nces)

2. *Alternaria brassicicola* (Schw,) Wiltshire

3. *Alternaria longissima*, Deighton Mae Garive

4. *Alternaria tenuissima* (Kunze.Fr.) Wiltshire

5. *Aspergillus candidus* Link

6. *Aspergillus fiavus* link ex. Fr.

7. *Aspergillus fumigatus*, Fresenius

8. *Aspergillus nidulans*, (Eidam.) Winter

9. *Aspergillus niger,* Van Tiegh

10. *Aspergillus oryzae* Ahlburg. Cohn

11. *Aspergillus ruber*

12. *Aspergillus sydowii* (Bainier and Sartory), Thom and Church

13. *Aspergillus tamarri,* Kita

14. *Aspergillus terreus*, Thom

15. *Botryodiplodia theobromae*, Patouillard

16. *Botrytis cinerea,* Persoon

17. *Cephalosporium humicola,* Oudemans

18. *Chaetomium brasiliense,* Bat and Pontuel

19. *Chaetomium globosum,* Kunze

20. *Cladosporium cladosporioides* (Fr.) de Vries

21. *Colletotrichum dematium* (Fr.) Grove

22. *Corynespora cassiicola* (Berk. and Curt.) Wei

23. *Cunninghamella elegans* Lendner

24. *Curvularia lunata* (Walker) Boediju

25. *Drechslera rostrata*

26. *Drechslera tetramera* (Mc.Kinney) Subram and Jain Syn. *Helminthosporium tetramera* Mc. Kinney *Bipolaris tetramera* (Mc. Kinney) Shoem

27. *Epicoccum purpurascens,* Ehrenberg

28. *Fusarium bulbigenum,* Cooke and Massee

29. *Fusarium equiseti* (Corda) Saccardo

30. *Fusariurn moniliforme,* Sheldon

31. *Fusarium oxysporurn,* Schlechtendahl

32. *Fusarium semitectum* Berkeley and Ravene

33. *Fusarium solani* (Martius)

34. *Fusarium udum* (Berkeley.)

35. *Macrophomina phaseolina* (Tassi.) Goid

36. *Memnoniella echinata* (Rivolta.) Galloway

37. *Mucor echinulatus*

38. *Mucor hiemalis*, Wehmer.

39. *Mucor varians*, Povah.

40. *Myrothecium roridum*, Tode.

41. *Nigrospora oryzae*, Hudson

42. *Penicillium corylophilum*, Dierck Syn. *P. umbonatum* Shopp.

43. *Penicillium chrysogenum*, Thom.

44. *Penicillium citrinum*,

45. *Penicillium expansum* (Link.) Thom.

46. *Penicillium oxalicum*, Thom.

47. *Penicillium rubrum,* Stoll

48.*Phoma humicola*, Gilman and Abbott

49. *Rhizoctonia bataticola* (Taub.) Butl.

50. *Rhizoctonia solani,* Kuhn

51. *Rhizopus arrhizus,* Fischer

52. *Rhizopus nigricans*, Ehrenberg

53. *Sclerotium rolfsii* Saccardo.

54. *Sclerotinia sclerotiorum*.

55. *Sclerotinia sclerotiorum*.

56. Sterile mycelium

57. *Verticillium albo-atrum* Reinke and Berthold

Fungi: Identification Character

.1. *Alternaria alternata* (Fries.) Keissler (Syn. *A. tenuis* Nces)

Colonies on Potato Dextrose Agar medium fast growing, usually black or olivaceous black at maturity, sporulation abundant; mycelium septate; branched; hyaline; late turning to dark olive buff; 2.90-4.90 μ in width; sometimes swollen (10.50 21.80 μ) to form chain of chlamydospores; terminal and intercalary; dark olive buff; measuring 12.50-23.40 μ in diameter, conidiospores arising singly or in groups; usually simple; septate; straight or variously curved; geniculate; pale to mild olivaeous brown; smooth, 21.90-65.40 μ in length and 2.80-7.29 μ in width; conidia formed in long and often branched chains; obclavate obpyriform; cylindrical; muriform, olive brown to dark brown or almost black; smooth; sometimes verrucose; multiseptate; with 1-6 cross walls and 0-4 longitudinal septa; measuring 15.30-42.80 x 7.0-13.50 μ in size; beaks olive buff in colour, 3.0-17.90 μ in length and 3.20-5.40 μ in width 0-12 septate.

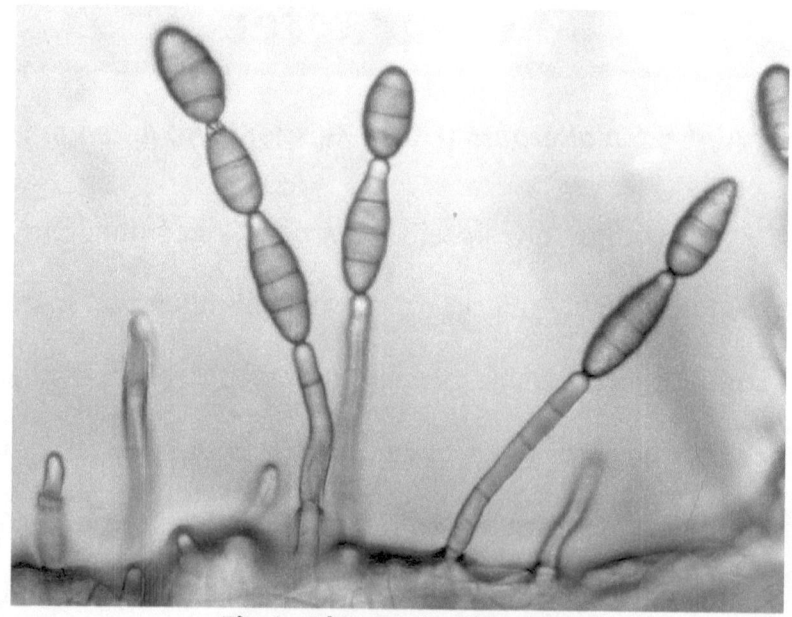

Fig.1. *Alternaria alternata*

2. *Alternaria brassicicola* (Schw,) Wiltshire

Colonies fast growing on Potato Dextrose Agar medium; dark brown to almost black; circular; zonate; vegetative mycelium hyaline at first, later brown or olivaceous; measuring 1.80-7.40 μ in width; conidiophores amphigenous; arising singly or in groups of 2-12; 0-3 septate; basal cell swollened; olivaceous; rarely branched; straight and upright when solitary; of tern curled when fasiculate; not markedly geniculate; usually with a terminal scar; sometimes with lateral scar also; measuring 22.50-70.0 x 5.20 -7.90 μ in size, conidia borned in chains of 20 or more; light olivaceous to dark olivaceous cylindrical; oblong; usually tapering slightly towards the apex or obclavate; with the basal cell rounded; beak usually nonexistent; apical cell rectangular age; measuring 18.50-128.75 x 8.60-29.70 μ; beak about one sixth of the length of spore and 6.50-8.0 μ wide, with 1-11 but usually less than 6 transverse septa and 6 longitudinal septa; a central pore visible in cross walls.

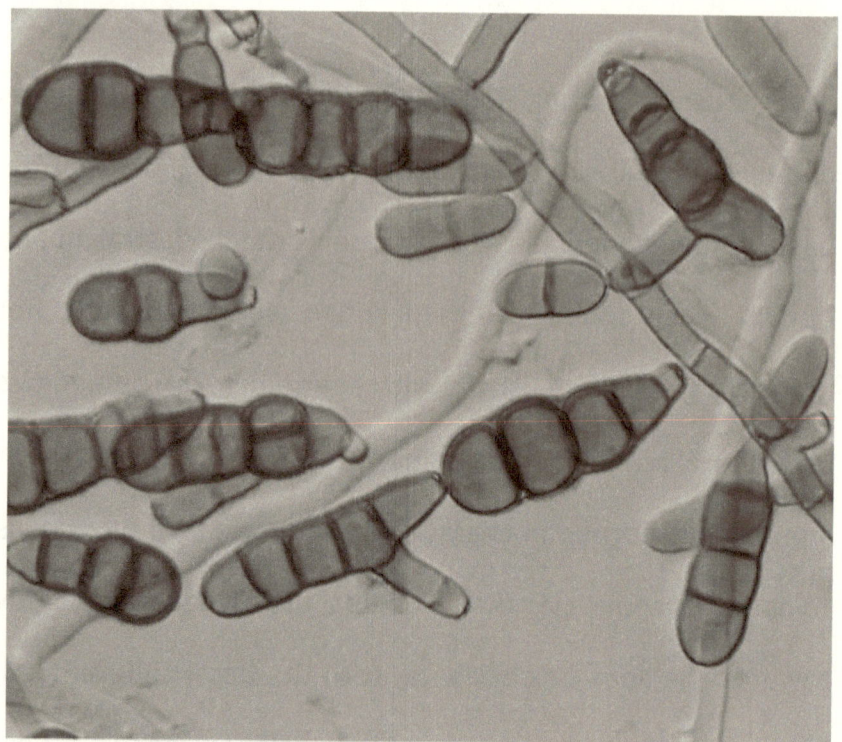

Fig.2. *Alternaria brassicicola*

3. *Alternaria longissima*, Deighton Mae Garive

Colonies fast growing on Potato Dextrose Agar medium; amphigenous; fluffy or hairy; dull; white to light brown; mycelium partly superficial and partly immersed; conidiophores erect or ascending; simple or occasionally branched; straight or slightly flexuous; sometimes geniculate; somewhat swollen at the apex; septate; pale or pale brown; smooth below; verrucose sometimes below the apex; measuring 28.50-36.15 x 20.6-6.58 μ in size; with several conidial scars; conidia solitary or in short chains of 2 or 3; variable in shape and size; straw coloured to brown; mostly long; obclavate; cylindrical; sub-cylindrical portion of a few to several cells to a very long; narrow; septate; filiform beak; pale to light brown; 5-40 transverse septa; 25.0-945.0 μ in length; 6.0-15.0 μ in width; broadest part about 230 μ; thick at the apex; conidiophores short; variable in size and shape; often with a few longitudinal or oblique septa; conidia thin walled; smooth except around the base; verrucose; dark brown, multicellular; muriform; chlamydospores 18.0-42.0 x

16.30 - 34.25 µ in size.

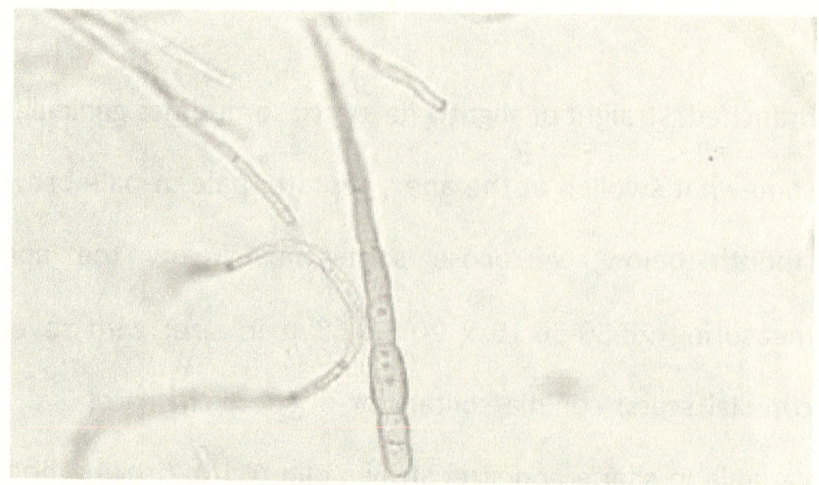

Fig.3. *Alternaria longissima*

4. *Alternaria tenuissima* (Kunze.Fr.) Wiltshire

Colonies on Potato Dextrose Agar medium fast growing; cottony smoky black; mycelium septate; branched; hyaline at first; later olive buff 280-8.30 µ in width; sometimes swollen (15.0- 20.14 µ) to form chains of chlaymydospores; conidiophores long, more or less cylindrical; simple or rarely branched; septate (2-11 cross septa); light to dark brown; 16.90-74.0 µ in length and 2.85-4.70 µ in width; conidia produced in short chains of 2-6; conical or oval dark olive; smooth; tapering gradually in beaks generally with 2-8 cross and 0-6 longitudinal septa; slightly or unconstricted at septa; 12.20-58.13 x 11.25- 18.0 µ in size; beaks almost equal to length of conidium; light in colour 11.35-55.25 µ in length and 3.25-5.12 µ in width with 0-8 cross septa.

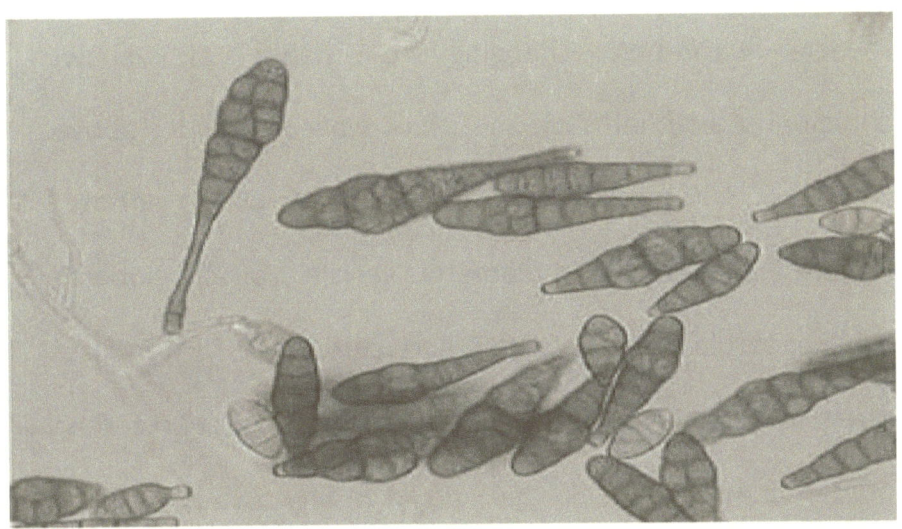

Fig.4. *Alternaria tenuissima*

5. *Aspergillus candidus* Link

Colonies on Czapek's agar medium moderately fast growing; white; changing to light buff mixed with pale; greenish; vinaceous; cream coloured; surface growth made up of conidiophores heads; vegetative mycelium submerged; reverse warm-buff changing to shades of yellow; conidiophore smooth; hyaline; thick walled; broader above 353.25- 858.30 x 5.60-9.39 µ in size; heads white; globose; radiate 25.0-305.64 µ in diameter; vesicle typically globose; hyaline; fertile over the whole surface 9.58-18.75 µ in size; rarely up to 50.0 µ in larger heads; sterigmata in two series; primary 5.47-8.64 x 1.80- 3.49 µ; secondary 6.15-7.6 x 2.35- 2.75 µ; conidia hyaline; globose with yellowish to pinkish buff envelope of hyphae bearing hull cells up to 27.0 µ in diameter; wall dark; reddish purple; made up on one layer of cells 137.46-361.85 µ; asci spored; release ascospores by breaking; ascospores purple red; lenticular; smooth walled with two equatorial crests 3.86-4.58 x 3.54-3.85 µ.

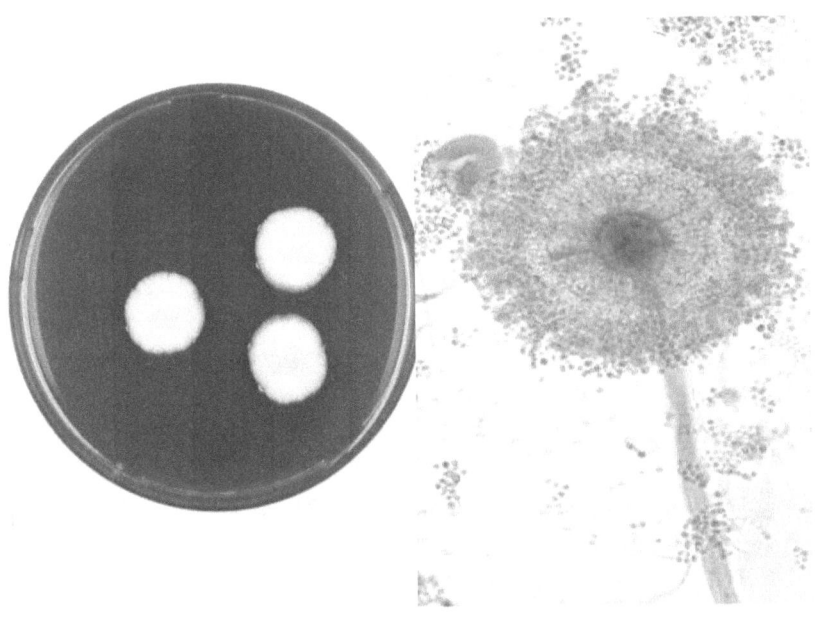

Fig.5. *Aspergillus candidus*

6. *Aspergillus fiavus* link ex. Fr.

Colonies fast growing on Czapek's Agar medium

spreading; somewhat floccose; sporulation starts from centre at first; white in the beginning and later developed in different shades of green; rock-green to yellowish green but entirely covered by sporulation in old colonies; reverse of the colonies yellowish and later turn into brown shades, with age; conidial haads both small and large present in same colony; yellowish green; globose to dome shaped; radiate; 90.80 μ in diameter: conidiophores colourless; rough walled 475.90-1479.30 x 6.40-9.50 μ in size; arising from submerged mycelium with prominent foot cell; continuous or septate narrow at base; 4.80-10.20 μ in width; gradually broadening to 10.45- 11.92 μ in middle and at apex 9.76-14.70 μ enlarging into dome like vesicle in smaller heads with limited fertile area: flask shaped in bigger ones with its entire surface Sterile; 15.48-66.29 μ in diameter; sterigmata in one series in small heads measuring 7.10-10.80 x 2.50-4.60 μ or in two series in larger heads; primary sterigmata 4.40-9.40 x 2.10-3.0 μ and secondary 6.30-9.20 x 2.20-3.0 μ; conidia almost globose to sub-globose or even pyriform; 2.90-4.84 μ in diameter; colourless to yellowish-green; echinulated; rough in appearance; sclerotia and perithecia not observed.

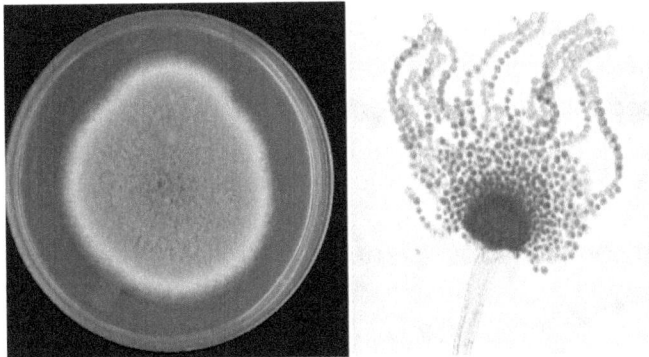

Fig.6. *Aspergillus fiavus*

7. *Aspergillus fumigatus*, Fresenius

Colonies on Czapck's Agar medium rapidly growing: velvety; rarely floccose; continued to scanty growth of few aerial hyphae; at first white; later changing to blue green to dark green; becoming black in age; spreading; reverse light buff becoming pale buff conidiophores short; smooth; often crowded green; arising directly from submerged hyphae or as branches form aerial hyphae; usually a septate; gradually enlarging upward with apical vesicles; 60.53-392.45 x 2.36-9.15 μ in size; conidial heads compact; columnar; green varying greatly in size in same culture up to 284.40-395.27 x

38.10-50.40 μ in size; xommonly much shorter; vesicle flask

shaped; fertile usually on the upper half; greenish 5.30-8.35 x

2.50-3.20 μ; phialides in one series; sterigmata more or less

bottle shaped; somewhat greenish yellow; closely crowded;

closely packed with axis; parallel to axis of stalk; 5.42-9.16 x

1.64-2.48 μ; conidia globose; rough to echinulate, dark green

in groups; light green; singly 2.37-3.29 μ in diameter.

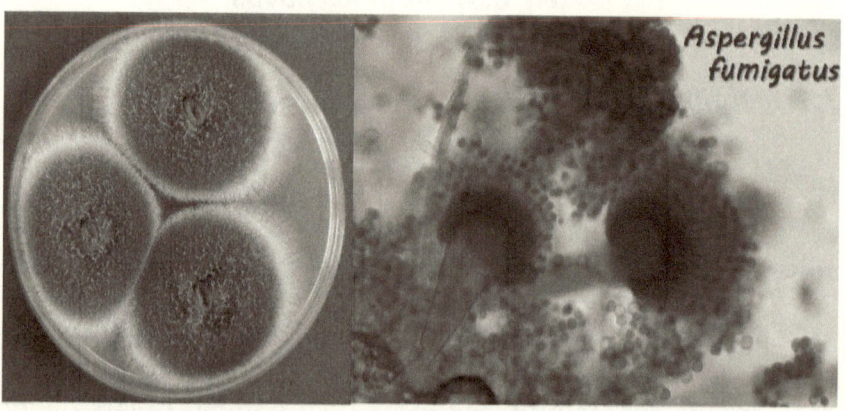

Fig.7. *Aspergillus fumigatus*,

8. *Aspergillus nidulans*, (Eidam.) Winter

Colonies on Czapek's Agar medium spreading broadly

with little floccosity; confined to centre only; white in earlier

stages; changing into smoky-grey with green buff margin with age; reverse cinnamon to pinkish cinnamon in shades of purple red; green heads on the centre of the colony; conidial fructification predominating during first two weeks; conidial heads short; green; columnar; long; 48.25-79.4 x 28.30-40.39 μ; occasionally up to 130.25 μ or even longer; conidiophores aerial: emerge directly from the submerged mycelium; smooth walled; more or less pale brown or in lighter shades of cinnamon brown: gradually broadening above; ranging from 65.47-128.0 x 2.75.3.49 μ; occasionally upto 180.94 μ in length: 2.76-5.38 μ near foot increasing to 3.89-6.45 μ below vesicle: hemispherical: more or less yellowish brown; 8.94-13.15 μ; occasionally upto 18.46 μ in diameter: phialides in two series; primary bottle shaped: yellowish brown; 6.45-6.96 x 2.17 x 3.65 μ; secondary 4.96-6.40 x 1.75-2.85 μ in size; conidia smooth; globose; pale yellow green to light yellow green; green in mass 2.38-3.75 μ in diameter; cleistothecia abundant; large; develop later from the centre of the colony; outwards olive buff to pale dull grey; reverse in shades of purplish red; becoming dark in age; 116.50-175.48 μ in diameter surrounded by hyphae and hull cells; 12.75-21.80 μ in diameter free; ascospores purple red; smooth walled; lenticular with two equatorial crests; 3.58-4.57 x 2.90-4.15 μ in diameter.

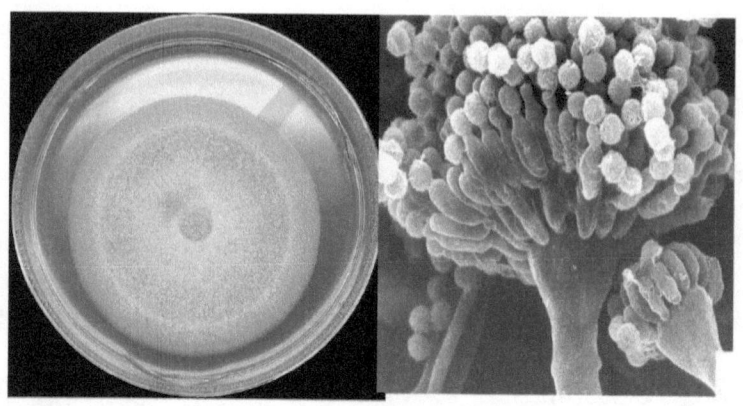

Fig.8. *Aspergillus nidulans*

9. *Aspergillus niger,* Van Tiegh

Colonies on Czapek's Agar medium growing fast with abundant submerged mycelium; aerial hyphae scantly produced; more or less yellow coloured, blackish brown in shades of carbonaceons black; reverse pale olive buff; conidial heads typically globose or radiate; blackish brown to purple brown in shades of carbonaceous black; periphery splitted into radiating columns of conidia; 630.70-918.40 μ in diameter; conidiophores often arising directly from the substratum; smooth; septate or aseptate; usually light brown near the vesicle; the remaining portion hyaline; varying greatly in length and diameter in same colony; commonly 359.35-1264.46 x 7.38-12.44 μ; rarely up to 10.0 mm. long and 20.0 μ in diameter; vesicle globose or sub-globose; dark yellowish brown; fertile over the entire surface; commonly 21.50- 51.20 μ in diameter; rarely upto 83.26 in diameter; phialides in two series; more or less brown coloured; primary sterigmata more or less wedge shaped; closely packed; varying greatly in length in same colony usually 19.70-26.85μ long and 8.60-4.50 μ in diameter; secondary sterigmata bottle shaped; uniform in diamensions; usually 6.50-9.50 x 2.30-

3.15μ; conidia globose: at first smooth; later echinulate, brown in mass 3.94-4.05 μ in diameter; rarely 5.0 μ sclerotia not observed.

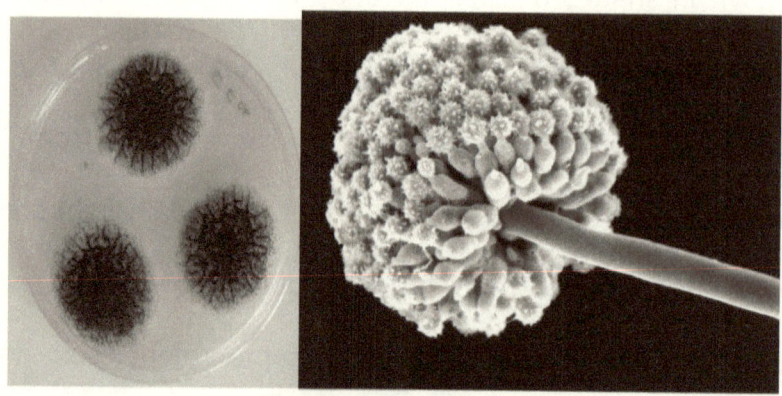

Fig. 9. *Aspergillus niger*

10. *Aspergillus oryzae* Ahlburg. Cohn

Colonies on Czapek's Agar medium fastly growing; first white; later changing to pale greenish yellow with the development of conidial heads; varying from yellowish green to dark green; reverse colour pale orange yellow; conidial heads abundant: radiate; dome shaped; 79.48-135.35 μ in diameter; conidiophores aerial emerging from the submerged mycelium; hyaline; longpitted: appearing rough in dry months: 676.30-1243.0 x 7.80-16.42 μ; vesicle globose to sub globose; hyaline; fertile over the whole surface 25.74-47.30 μ in diameter with walls; sterigmata in one series; bottle shaped; yellowish- green; densely crowded 6.45-13.50 x 3.32-4.50 μ; conidia pyriform to globose; hyaline; yellowish green in masses; 3.40-5.26 in diameter.

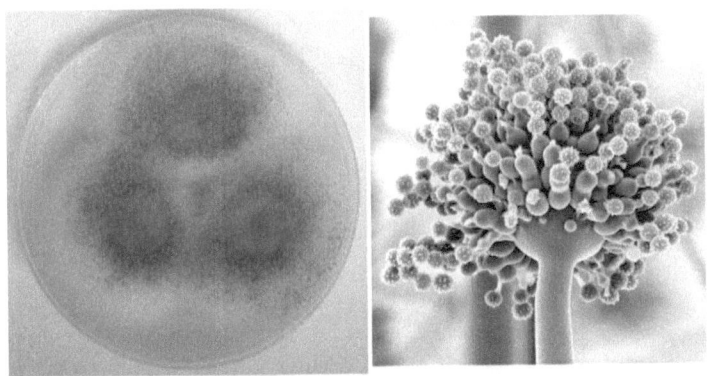

Fig.10. *Aspergillus oryzae*

11. *Aspergillus ruber*

Colonies on Czapek's agar medium fast growing and regular; plane; predominantly red; ranging from ferruginous to morocco-red; conidial heads above the felt pale grey to green to deep olive-grey; numerous; generally crowded near the centre of the colony; reverse in shades of dark red-brown;

radiate; 160.25-249.58 µ in diameter; conidiophores emerge directly from the substratum; smooth hyaline to orange brown; 512.30-750.0 µ in length; terminating into vesicle; 26.0-34.0 µ in diameter; phialides in a single series; 7.0-8.70 µ.; conidia; elliptical to sub-globose; closely spinulose 5.0-5.80 µ in long axis; cleistothecia abundant in dense layer at agar surface; yellow to orange-red; spherical to sub-spherical; 84.50-119.80 µ in diameter; asci 12.0-14.70 µ.; ascospores lenticular 5.2-6.0 x 4.40-4.75 µ with furrow in form of a borad depression around equator; ridges low; walls smooth except along ridges.

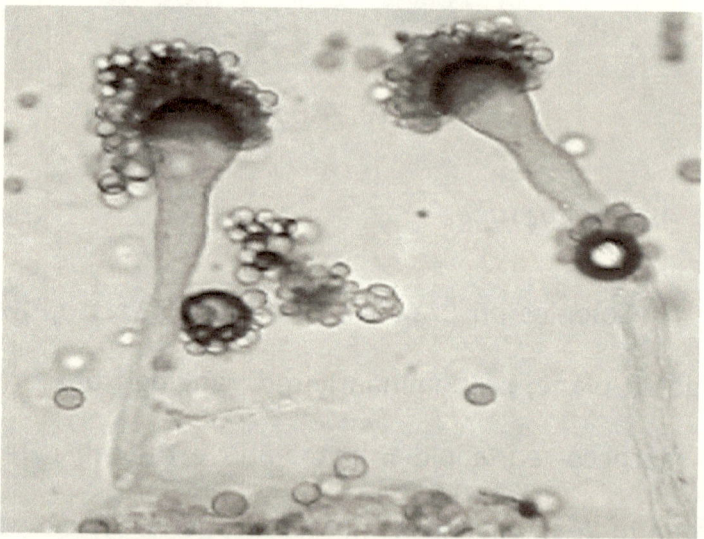

Fig.11. *Aspergillus ruber*

12. *Aspergillus sydowii* (Bainier and Sartory), Thom and Church

Colonies on Czapek's agar medium fastly growing at room temperature; more or less floccose; strongly felty; close textured with trailing and interlacing hyphae; initially white; changing to pale grey with age, vinaceous when old; reverse orange-red to almost black; conidial area velvety; bluish green with bluish effect; specially prominent in young fruiting portion; becoming light cinnamon dark to dark in age; reverse warty; ocher red to marron; conidial heads radiate or globose; leafy green to light green; 16.54-36.74 μ in diameter; conidiophores aerial; mostly arisefrom submerged gypae; colourless; smooth; thick walled; long pitted; appearing rough in dry months; measuring 684.5-1218.0 μ in length and 7.12-16.3 7 μ in breadth; vesicle hyaline; globose to sub-globose; fertile over the entire surface, 7.35-18 μ in diameter; phialides in two series; colourless; primary 5.72-6.46 x 2.34-2.86 μ; secondary 6.80-10.48 x 1.46-2.30 μ; conidia globose;

spinulose; more or less green in mass, 2.15-35.87 μ in diameter.

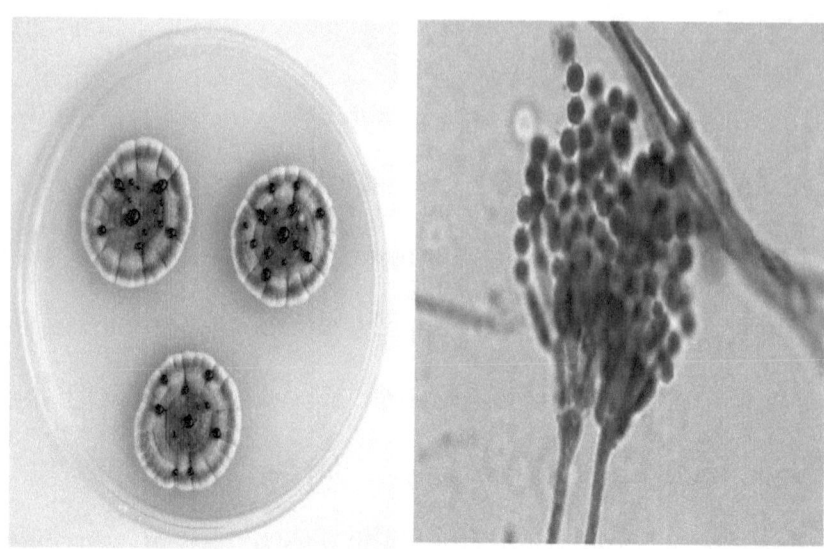

Fig.12. *Aspergillus sydowii*

13. *Aspergillus tamarri,* Kita

Colonies on Czapek's agar medium spreading broadly with vegetative hyphae; mostly submerged with fruiting areas; at first colourless passing through orange-yellow shades to brown in old colonies; light brownish olive; medal bronze and raw umber; reverse uncoloured or occasionally pinkish; conidiophores arising from submerged hyphae; 1.0-

2.0 mm. in length; 9.48-   18.50 μ in diameter; increasing in diameter towards the apex passing into vesicles; vesicles 24.40-25.14 μ in diameter; conidial heads vary greatly in size in the same fruiting area; globose to columnar with radiating chains and columns of conidia; phialides in one series in small heads; in two series in large heads; primary 7.40-9.56 x 2.80-3.75 μ; secondary 7.36-9.85 x 1.95-2.80 μ; conidia more or less pyriform to globose. 5.0 μ or 6.0 μ or 8.0 μ in diameter; rough; sclerotia occasionally produced; usually purple or reddish purple; globose or pyriform with white apex.

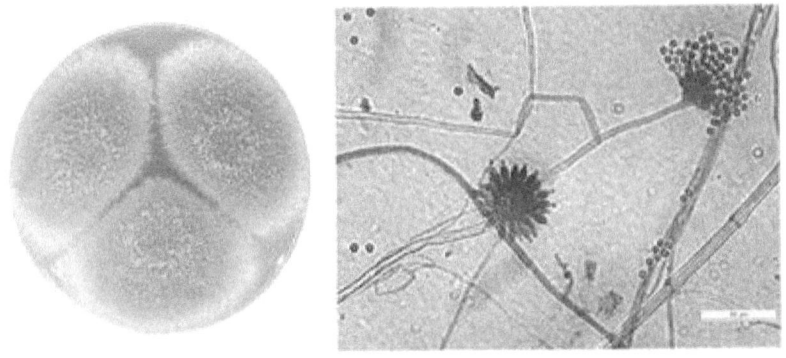

Fig.13. *Aspergillus tamarri*

14. *Aspergillus terreus*, Thom

Colonies on Czapek's agar medium fast growing; velvety; initially white; changing through yellowish shades to various cinnamon lints and ultimately wood brown to avellaneous; reverse warm buff passing through deeper yellow shades to sudan-brown or Brussels brown; conidiophores hyaline; smooth; more or less flexuous; approximately uniform in width throughout, commonly 148.36-235.49 x 3.44-5.68 μ; conidial heads long; columnar; almost uniform in diameter usually 145.35-492.67 x 33.40-51.60 μ; vesicle sub-globose to dome shaped; hyaline, 10.48-16.49 μ; occasionally upto 18.95 μ in diameter; phialides in two series; hyaline; closely packed; primary sterigmata 4.83-5.78 x 1.6-2.73 μ; secondary sterigmata 5.74 x 6.90 x 1.54-19.86 μ; conidia usually globose; smooth; 1.3-2.85 μ.

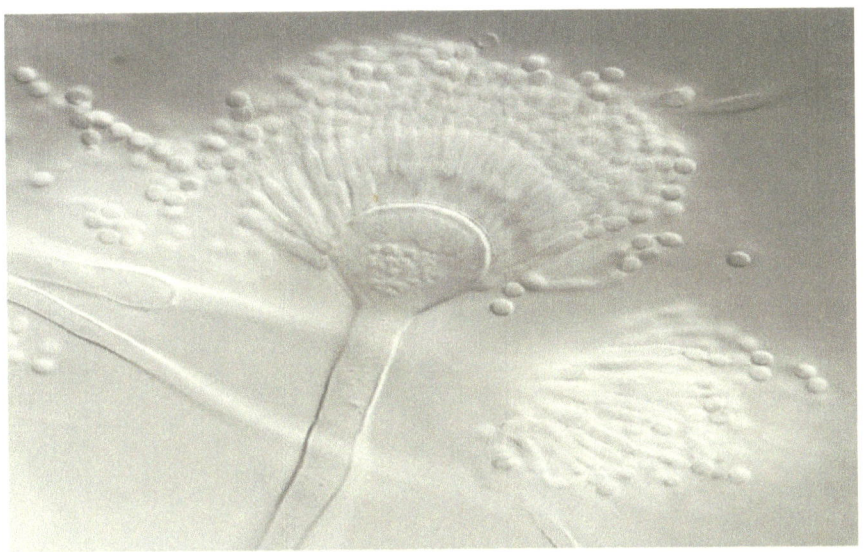

Fig.14. *Aspergillus terreus*

15. *Botryodiplodia theobromae*, Patouillard

Colonies on Potato Dextrose agar medium moderately growing; circular; moderately floccose; white in beginning; changing reddish brown to dark pinkish; mycelium septate; branched; pale brown; varying from 1.40-3.29 μ in diameter; pycnidia caespitose or botryose; more or less stromate; erumpent; walls membranous; carbonaceous black; globose; often papillate on a hairy stroma; 158.42-199.95 μ in diameter; conidiophores elongated; hyaline; 45.80-50.15 μ; conidia elongated; slightly brownish; one septate; 25.80-

33.70 x 1295-14.92 μ in size.

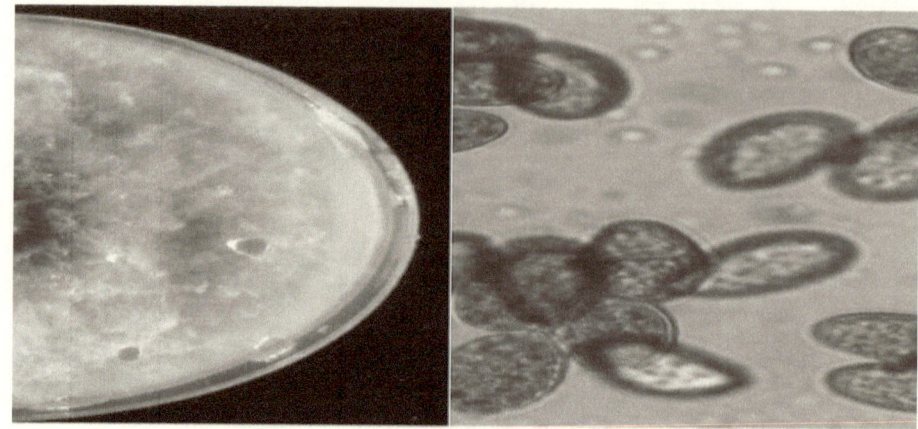

Fig.15. *Botryodiplodia theobromae*

16. *Botrytis cinerea,* Persoon

Colonies fast growing on Potato Dextrose agar medium with aerial mycelium diffuse; grey; grey-green; dark olive-green to dark brown to almost black or reddish green; dusty; hyphae creeping; dense; 1.90-2.0 mm high; conidiophores erect; branched or unbranched; septate 11.60-20.32 μ thick; brown or almost black; towards the tip almost hyaline; narrowing to a point truncate or with swollen warts on the tips; where

the conidia are formed singly of fine warts; conidia are attached on the projections forming a thick head which soon falls off; conidia one-celled; ovate or elliptical

; or globose; finely apiculate at the base; measuring 8.50-10.8

x 6.30-10.10 µ; hyaline or bright coloured with a thick

brownish wall.

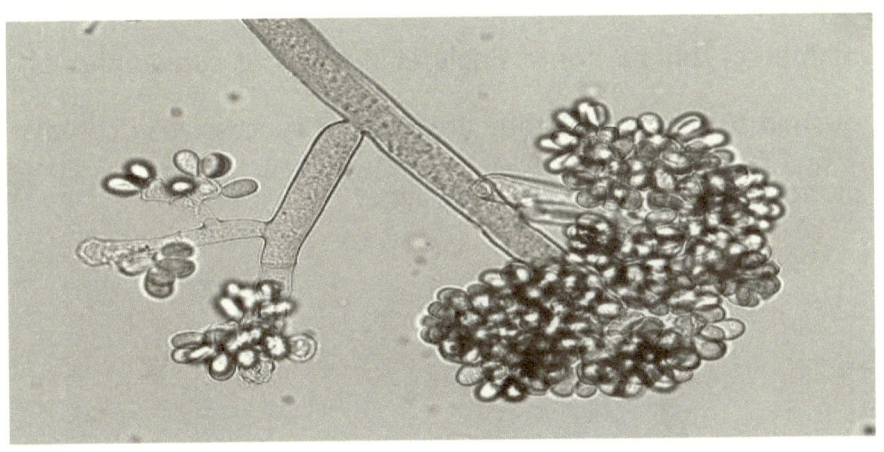

Fig.16. *Botrytis cinerea*

17. *Cephalosporium humicola,* Oudemans

Colonies on Potato Dextrose agar and Czapek's agar medium fast growing spreading; orbicular; floccose; at first white; later with white margin and dilute rose coloured centre; Sterile hyphae septate; branched; hyaline; creeping 3.81-4.90 µ thick; intermingled with segments, which appear like chlamydospores; conidiophore arise as short branches of aerial hyphae; erect; non-septate, non-swollen at the tip; 105.30-197.80 µ long; conidial heads 22.38-24.27 µ in diameter; conidia borne singly at the tips of conidiophores; pushed to the side at they are formed successively; conidia globose: 2.35-2.50 µ; almost hyaline.

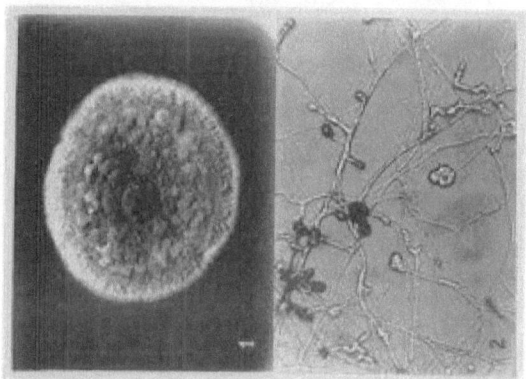

Fig.17. *Cephalosporium humicola*

18. *Chaetomium brasiliense,* **Bat and Pontuel**

Colonies on Potato Dextrose Agar medium moderately fast growing; white; floccose; perithecia developing after a week as superficial gregarious or scattered; ovate to ellipsoidal; ostiolatel olivaceous to dark brown in colour; clothed with hairs on all sides except at the base, 80.8-119.0 x 65.70-101.90 μ in size. Hairs simple; broador at base and narrower towards apex.; sparingly septate; slender; straight to flexuous; dark brownish black colour; asci oblong; clavate; 8-spored; ascospores ellipsoidal usually; rarely oval; single celled; distinctly apiculate at ends olivaceous brown; smooth walled; usually 4.30-15 x 3.08-5.12 μ in size.

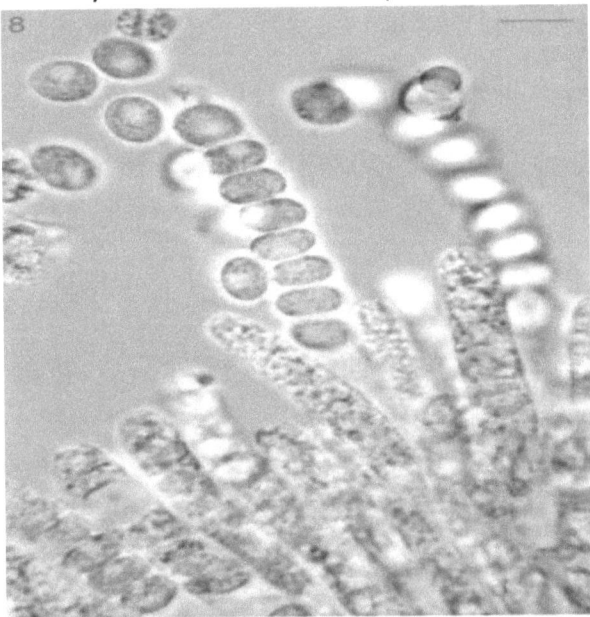

Fig.18. *Chaetomium brasiliense*

19. *Chaetomium globosum*, Kunze

Colonies white to milky white on Potato Dextrose Agar medium: olivaceous in young stage: later turns dark brown in dry specimens; reverse hyaline: perithecia scattered or gregarious: broadly ovate or ellipsoid: pointed at base; thick covered with slender and flexous hairs with a well defined ostiole: measuring 258.40-301.87 x 205.60 µ; apical hairs simple; some what coarser; sparingly septate; minutely scabrous 3.0-4.0 µ thick; 700.0 µ long in fresh condition; pale yellow; light brown in dry condition; measuring 327.50-613.80 x 2.75-4.30 µ in size; asci oblong-clavate; 8 spored; evanescent; ascospores dark; small; broadly ovoid; sometime at both the ends; 7.90-9.36 x 5.94-7.35 µ in size.

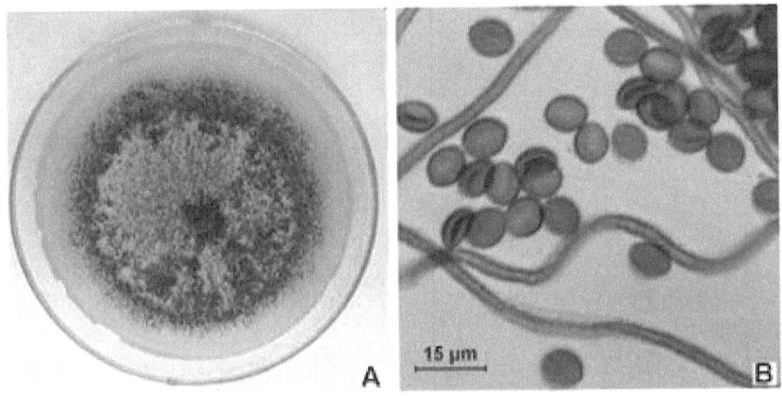

Fig.19. *Chaetomium globosum,*

20. *Cladosporium cladosporioides* (Fr.) de Vries

Colonies on Potato Dextrose Agar medium usually dark green; velvety and relatively slow growing; aerial mycelium rather scanty but normally sporulate freely; mycelium composed of narrow hyaline hyphae: 2.0-4.0 µ in width; dark hyphae 5.0-11.80 µ in width; conidial heads composed of large number of smooth walled spores; a small proportion of which are large that the remainder; conidia formed in chains; reverse of the colony dark coloured; mycelium hyaline; becoming dark; septate, smooth or finely roughend; 2.0-3.50 µ thick; conidiospores arise laterally from the mycelium or formed terminally on the hyphae; brown; smooth or finely roughened; usually unbranehed; 27.50-82.40 μ long; 2.90-5.40 μ in diameter; septate without prolongation and inflations; constricted at septa; canker and more uniform in diameter than the vegetative hyphae; often with two short and lateral outgrowths; just beneath the septum bearing chain of conidia; conidia born in chains; brown; variable in shape; continuous; occasionally 1 or 2 septate; not obviously rough walled; sometimes tending to become cylindrical; pointed towards the base and irregular at the distal end due to protrusions on which conidia formed, measuring 3.70-15.80 x 2.0-3.90 μ in size.

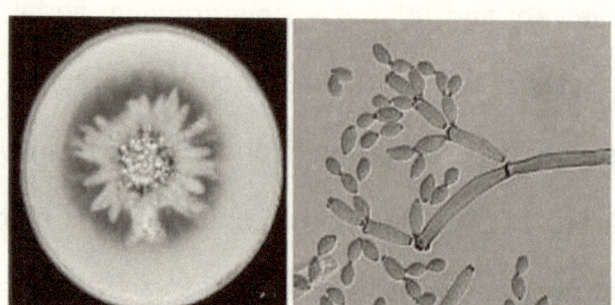

Fig.20. *Cladosporium cladosporioides*

21. *Colletotrichum dematium* (Fr.) Grove

Colonies on Potato Dextrose Agar medium compact; creamy in colour in initial stage which later on turn blackish brown to dark brown; reverse of the colony black; mycelium slender; septate; when young gyaline, containing droplets of oil globules and later become brownish in colour; 3.50-4.90 μ in thickness; acervuli single or in groups; hemispherical; immature acervuli grayish-white to dull orange and mature dark brown in colour; setae numerous; trichiform, brown to blackish brown; longer than conidial mass, 2-7 septate; 75.20-214.50 x 3.0.-4.80 μ and varied in number from 2-15 per acervulus; conidial mass white to dull white: pale orange or bright orange: conidiophores arranged in different layers on the surface of stromatic tissues; short; erect; simple; hyaline; aseptae and packed together; measuring 12.80-19.40 x 3.0-4.30 μ; conidia gyaline; when single but white to dull white or pale orange in mass; fusoid; ends rounded or slightly tapering measuring 19.28-25.39 x 3.0-5.0 μ in size.

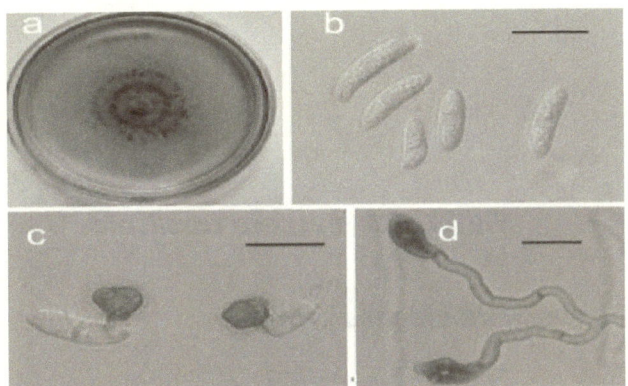

Fig.21. *Colletotrichum dematium*

22. *Corynespora cassiicola* (Berk. and Curt.) Wei

Colonies on Potato Agar medium moderately growing, sprading, sub-floccose; circular, effused, grey to brown; mycelium simple or branched; septate; sub-hyaline to pale brown; 1.93-7.25 µ with; Conidiophores straight or slightly flexuous; pale to dark brown; with 2-9 cylindrical proliferations; 35.0-400.0 µ in the length; 4.95-9.52 µ in width; conidia formed singly or in chains 2-6 at apex of proliferating conidiophores; obclavate or cylindrical; straight or curved; smooth; sub-hyaline to pale olivaceous brown with 4-18 pseudosepta 13.94- 144.25 x 4.80-11.92 µ in size.

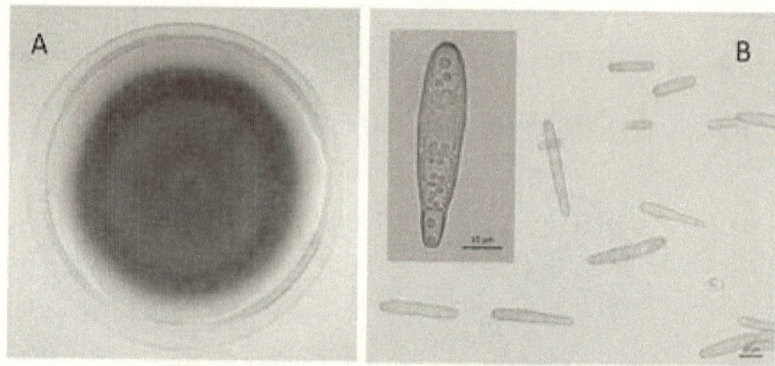

Fig.22. *Corynespora cassiicola*

23. *Cunninghamella elegans* Lendner

Colonies fast growing on Potato Dextrose Agar

medium; turf white to silver, spreading; mycelium white, floccose; filaments firm and interwoven 7.50-13.0 μ wide; with abundance of oil; circinate portions typical; continuous, when young; later becoming septate; conidiophores erect; multibranched; terminal vesicles 29.35-34.95 μ in diameter; spherical; smooth; lateral branches lacking or up to 3 whorled; place of attachment of conidiophores swollened; sub-terminal whorl 38.0 μ long; vesicles spherical 18.50-27.40 μ in diameter; smooth; basal whorl of pyriform branches measuring 14.0-25.0 μ in length, smooth; super branches arising from tenninal head of varying lengths; terminal conidia lemon shaped; spicules present after separation from vesicles; measuring 13.50- 8.90 μ in width; fenely echinulate; lateral conidia ovate measuring 4.5-6.0 x 8.30-10.0 μ in size;

non-spiculate; finely echinulate.

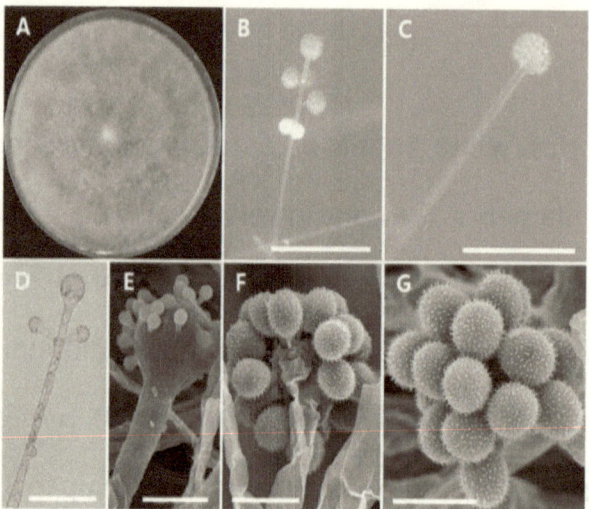

Fig. 23. *Cunninghamella elegans*

24. *Curvularia lunata* (Walker) Boediju

Colonies on Potato Dextrose Agar medium fast growing; sub-floccose, appressed; smooth; at first white becoming dark; greenish olive to black in colour due to abundant sporulation, reverse of the colony olivaceous black and smooth; hyphae septate; branched; light brown to olive brown 3.40-6.20 in width, conidiophores septate; erect; long; narrow; unbranched; geniculate at the tips; brown; measuring 19.50-107.08 x 3.85-4.91 μ in size: conidia present in a whorl at the tip of conidionhore: conidia mostly 3-septate; rarely 1-5 scptate; the third cell from base larger in size and darker in colour than other cells; pale brown; clavate or pear shaped; straight or curved, echinulated; constrictions more prominent in septa; measuring 14.50-30.18 x 6.30-11.20 µ in size; chlamydospores spherical; dark brown to black; measuring 10.90-17.40 µ in diameter.

Fig. **24.** *Curvularia lunata*

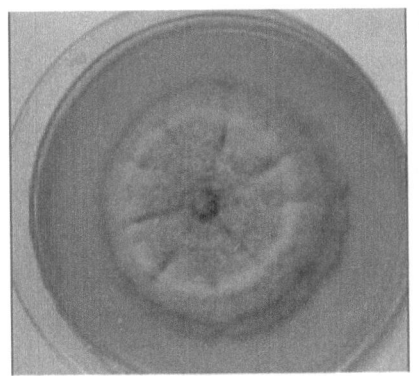 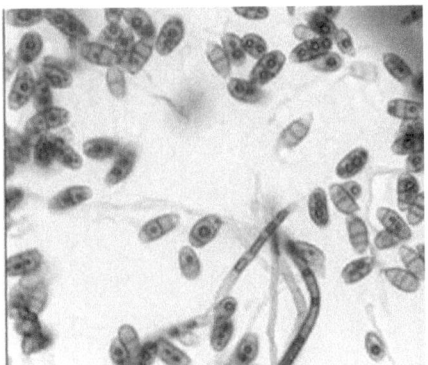

25. *Drechslera rostrata*

Colonies on Potato Dextrose Agar medium fast growing; velvety, grey in beginning but becoming grey brown with age; mycelium septate; pale-brown, varying from 3.69-4.53 μ in width; conidiophores solitary; straight of flexuous; geniculate; pale to brown in colour; up to 195.0 μ in length; conidia curved; fusiform to broadly ellipsoidal; dark; olivaceous brown; smooth; 5-10 septate; measuring 43.80-90.3 μ in length.

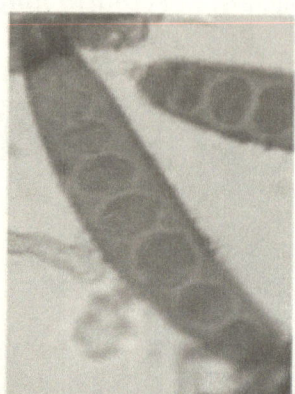

Fig.25. *Drechslera rostrata*

26. *Drechslera tetramera* **(Mc.Kinney) Subram and Jain Syn.** *Helminthosporium tetramera* **Mc. Kinney** *Bipolaris* *tetramera* **(Mc. Kinney) Shoem.**

Colonies on Potato Dextrose Agar medium fast growing; compact; sub-floccose; becoming more or less felty and plane later on; deep quaker drab to dark grayish olive in colour; reverse pale olive to light brown; mycelium septate branched; dark olivaceous; measuring 3.64-4.79 µ in width; conidiophores arise singly or in clusters of 2-3 bearing conidia acropleurogenously at the tip., simple; septate; brown; straight at base but geniculate; irregularly at tips; measuring 45.60-175.90 x 6.40-8.20 µ in size; conidia mostly 4-celled; light to dark brown; ellipsoid to almost cylindrical with rounded ends; lighter towards terminal; measuring 18.50-35.20 x 6.80-11.40 µ in size.

Fig. 26. *Drechslera tetramera*

27. *Epicoccum purpurascens*, Ehrenberg

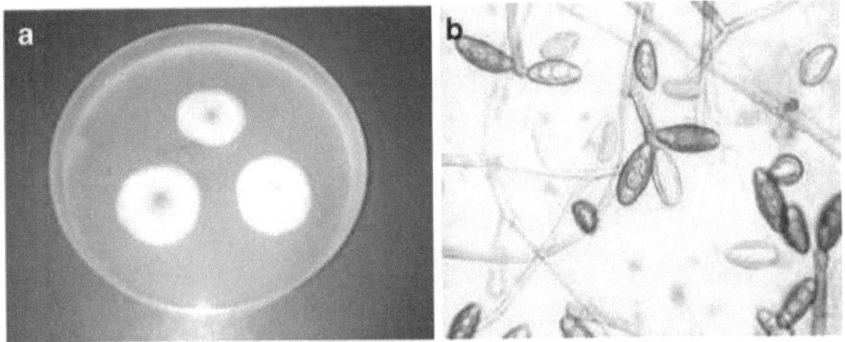

Colonies on Potato Dextrose Agar medium, moderately growing: develop a striking yellowish-green to purple red pigmentation on the underside; remain mostly Sterile; mycelium septate; branched; cream coloured; 2.85-3.49 µ in width; sporodochia scattered; punctuate; globose small; brown on a gemispherical somewhat depressed; black stroma; conidiophores arise from the stroma; club shaped; nonsepated; shor; dark coloured or black; measuring 12.0-13.80 x 5.10-6.80 µ in size; conidia present singly at the tips of conidiophores: globuse; scarcely stipitate at first yellow; later brown; finely warted; reticulately, verrucose with a tapering hyaline, measuring 16.40-21.90 µ in diameter.

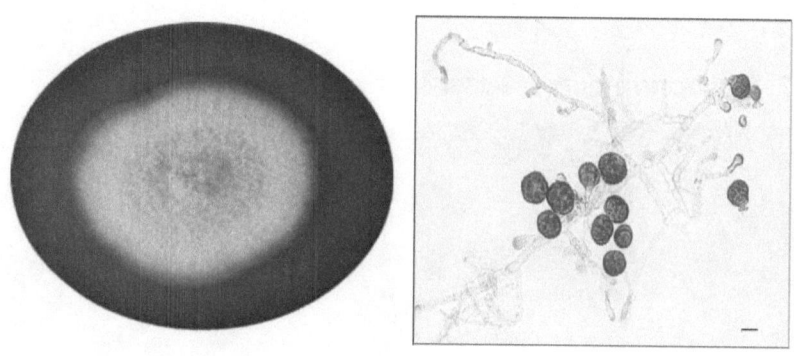

Fig. 27. *Epicoccum purpurascens*

28. *Fusarium bulbigenum*, Cooke and Massee

Colonies on Potato Dextrose Agar medium moderately

floccose; bright pink or lilac coloured aerial mycelium; sometimes rough; septate; varying from 1.49-5.80 μ in width; conidia in sporodochia on flat or round stroma; plectenchymatic; pale or rosy to red-violet; scattered over the substrate with dissolving orchre to lake-coloured; slimy conidial membranes and later developing numerous submerged or on the aerial mycelium; microconidia smaller; one celled or sparingly septate; macroconidia larger in sporodochia and pionnotes; 3-5 septate; long; awl-shaped; straight or slightly sickle-shaped; tapering at both the ends; some what constricted and hooked at the tip; tapering; the basal cell more or less pedicellate; measured on the basis of septation:

Number of septa	Size of microconidia and macroconidia
0	5.60-10.90 x 2.0-3.40 μ
1	12.80-31.25 x 2.0-3.60 μ
2	22.54-50.49 x 2.50-3.4 μ
5	34.90-65.80 x 3.20-4.30 μ

Chlamydospores terminal and intercalary both; pink or brown coloured; globose or ovate; smooth; one or two celled; in chains measuring 6.80-11.50 μ in diameter.

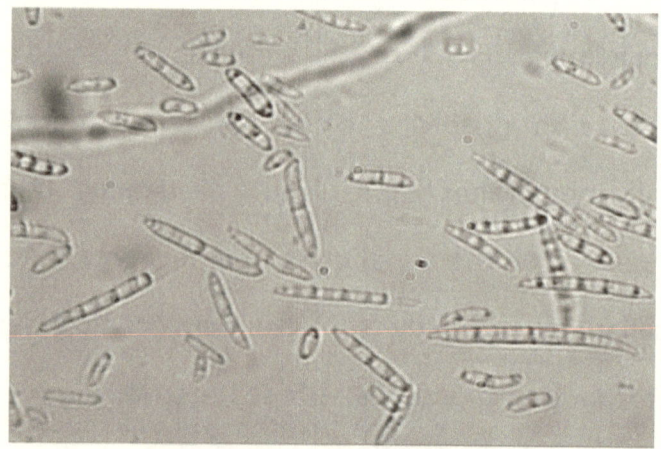

Fig. 28. *Fusarium bulbigenum*

29. *Fusarium equiseti* (Corda) Saccardo: Syn. *F. falcatum* Appel and Wollenweber *F. viticola* Thuemen

Colonies on Potato Dextrose Agar medium moderately floccose; cottony white; changing to pale brownish in colour; mycelium septate; hyaline; branched; measuring 1.58-6.25 in width; conidiophores simple or branched; compressed; bush like; in two or three or many times in multiple whorls; sterigma like papillae at the tips; bearing conidia at the tips;

Sterile and fertile hyphae septate; 3.85-5.49 μ in diameter in sporodochia; sporodochia pale brown; microconidia yellow to pink; one celled or 3-septate; oval; long; spindle or sickle shaped; macroconidia abundant in aerial mycelium in masses and also shaped; macroconidia abundant in aerial mycelium in masses and also forming sporodochia and pionnatal layer on agar surface; scattered as dust; pale yellow to white or drying cinnamon brown; brighter in colour as dry powder: spindle shaped; tapered at the ends; extended the tip into a thin straight or curved to parabolic or sickle shaped with a foot cell at the base; 4-11 septate: very rarely 12-13 septate: measured on the basis of septation:

Number of septa	Size of microconidia and macroconidia
0	4.82-6.02 x 3.05 μ
1	6.65-11.70 x 3.07-3.70 μ
2	9.94-18.30 x 3.07-3.65 μ
3	12.40-24.50 x 3.07-3.65 μ
4	25.59-40.73 x 3.69-4.92 μ

5	27.03-42.95 x 4.95 μ
6	35.48-46.83 x 4.95 μ
7	36.25-49.10 x 4.95 μ
8	37.93-51.43 x 4.95 μ
9	41.36-51.43 x 4.95 μ
10	40.51-52.38 x 4.95 μ
11	43.29-54.0 x 4.95 μ
12	46.0-65.50 x 4.95 μ
13	51.84-74.96 x 4.95 μ

Chlamydospores round; smooth; intercalary; terminal, many celled in chains or knots; 1-celled; brown coloured; measuring 6.80- 13.50 μ in diameter

Fig.29. *Fusarium equiseti*

30. *Fusariurn moniliforme,* **Sheldon**

Colony white on Potato Dextrose Agar medium; smooth; white or purple in colour; comact; powdery with production of microconidia; later become scattered or bright yellow to rosy white aerial mycelium as shining powder; mycelium septate; branched; white or purplish or very little orange white; varying from 1.60-4.30 μ in thickness; microconidia produced in long chains; dominating over macroconidia in false heads; small; 1-2 celled; spindle egg shaped; hyaline; measuring 9.60-26.10 x 2.35-4.50 μ in size; small heads of microcoiñdia on short and simple sterigma; macroconidia thin walled; scattered; narrowmg at both ends; constricted at tips; slightly curved; bright in mass; salmon coloured; on drying brick-red to cinnamon brown; formed in sporodochia or pionnotes; 1-5 septate; mostly 3-septte; seldom 6-7 septate; measured on the basis of septation:

Number of septa	Size of macroconidia
0	14.80-21.90 x 3.10-4.70 μ
1	26.40-37.50 x 4.0-5.20 μ
2	34.20-44.90 x 4.80-6.0 μ

Chlamydospores not present.

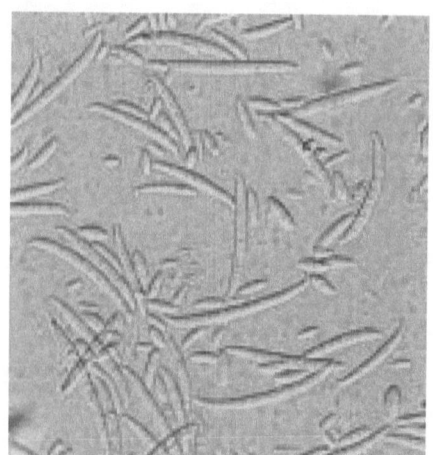 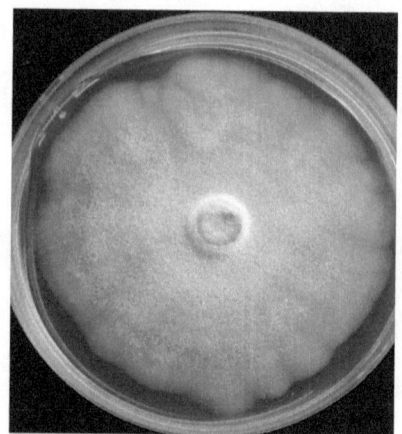

Fig.30. *Fusariurn moniliforme,*

31. *Fusarium oxysporurn*, Schlechtendahl

Colonies white in early stages: later changing to purple coloured; fast growing on Potato Dextrose Agar medium. floccose; stroma whitish brown to violet; smooth; hard; wrinkled under moist conditions; covered by fasicled aerial mycelium; later forming sporodochia; 0.60-2.50 μ in diameter; microcondia numerous; in groups; one or two celled; straight; curved or sickle shaped; one end pointed;

other obtuse; 0-septate; measuring 10.80-19.50 x 2.45-4.10 μ

in numerous in number; macroconidia in sporodochia oxhorn

shaped; 1-3 septate; commonly 1-3 septate; bluish pink

coloured: measured on the basis of septation:

Number of septa	Size of macroconidia
1	10.50-25.92 x 2.30-4.30 μ
3	19.80-44.70 x 2.80-5.0 μ
5	32.65-59.84 x 3.8-4.90 μ

Chalmydospores terminal and intercalary; globose;

smooth or wrinkled one celled, seldom two celled; brown or

dark brown in colour; measuring 8.35- 14.50 μ diameter.

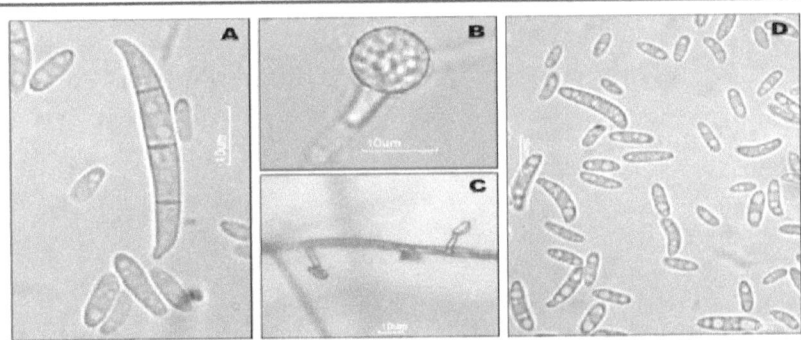

**Fig. 31 Morphological characteristics of cultures *of F. oxysporum*
isolated fromwilted watermelon. a Macroconidia. b Chlamydospore. c
False heads of microconidia formed on short monophialides. d
Microconidia**

32. *Fusarium semitectum* Berkeley and Ravene

Colonies fast growing on Potato Dextrose Agar medium; aerial mycelium; white-incarnate or isaoellin; with abundant aerial mycelium of whitish cottony growth in the earlier stages and later turned into light yellow or buff brown; mycelium septate 3.20-5.80 µ thick; creeping; sporodochia lacking; stroma plectenchymatic; bright brown; conidiophores arise singly or in groups; whitish; shiny highly branched bearing easily recognizable microcoidia giving an appearance of "Flower"; macroconidia on aerial mycelium as fusoid hyaline; scattered on aerial mycelium; 0-5 septate but occasionally upto 7-septate; wedge shaped measured on the basis of septation:

Number of septa	Size of macroconidia
0	4.30-15.80 x 2.0-3.80 µ
1	8.50-19.50 x 2.0-4.30 µ

0	23.50-49.60 x 3.20-6.10 μ
5	32.90-64.25 x 4.30-6.40 μ
7	39.20-76.50 x 4.50-6.30 μ

Microconidia and chlamydospores not observed

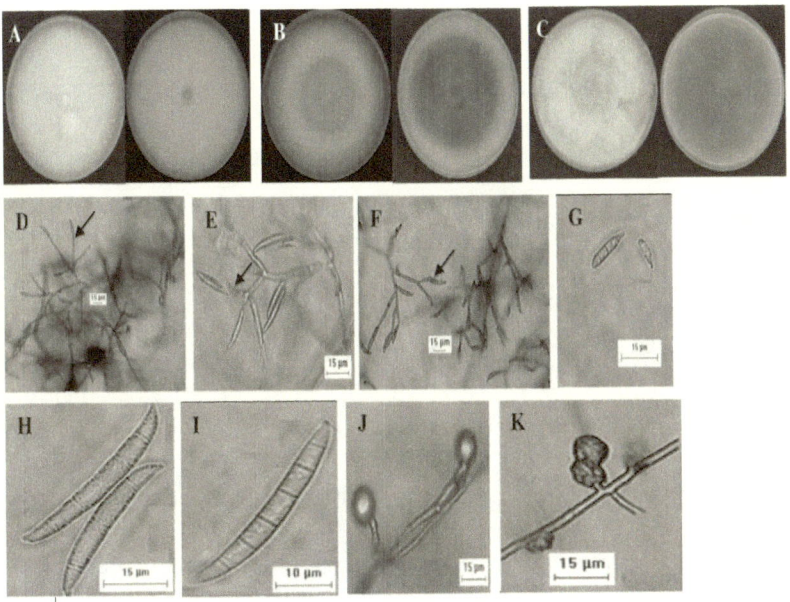

Fig.32. Colony colour, pigmentation, macroscopic and microscopic characteristics of *F. semitectum* isolates. A, White colony with pale cream pigmentation; B, White to brown colony colour with pale peach pigmentation; C, Brown colony with dark brown pigmentation; D-F, Mesoconidia with 'rabbit ear' appearances (arrow); G, Mesoconidia; H, Macroconidia with 3- to 5-septa; I, Macroconidia with 6-septa; J, Singly chlamydospores and K, Chlamydospores formed in pair.

33. *Fusarium solani* (Martius) Appel and Wollenweber. Syn. *F. alluviale*. Woolenweber and Reinking

Colonies moderate growing on Potato Dextrose Agar medium; grayish white; floccose with bluish to bluish brown; mycelium extensive; cottony; conidiophores variable in size; slender; simple; short; stout; branched irregularly bearing a whorl of conidia; single or grouped in sporodochia or pionnotes; whitish; olive yellow; often blue black held in mass of gelatinous material; microconidia abundant; hyaline oval to allantoid; 0-I celled; ovoid or belong, elongated; narrow at the apex; 0-septate measuring 4.90-13.20 x 2.35-4.55 μ I-septate; with measuring 23.40-24.50 x 3.20-4.50 μ; macroconidia sickle-shaped; 3-5 septate with a foot cell at the base; slightly curved; both end rounded to tempin like base; rare slightly pedicellate, measured on the basis of septation:

Number of septa	Size of macroconidia
3	20.38-49.50 x 3.80-7.0 μ
5	35.94-67.90 x 4.20-6.80 μ

Chlamydospores numerous; one or two celled; round

or oval; brown; single; globose to pear shaped; intercalary or terminal; thick walled; with smooth; sometimes finely warted when dry walls; measured on the basis of septation:

Number of septa	Size of chiamydospores
1 celled	8.20 x 8.0 μ
2 celled	9.35 x 6.10 μ

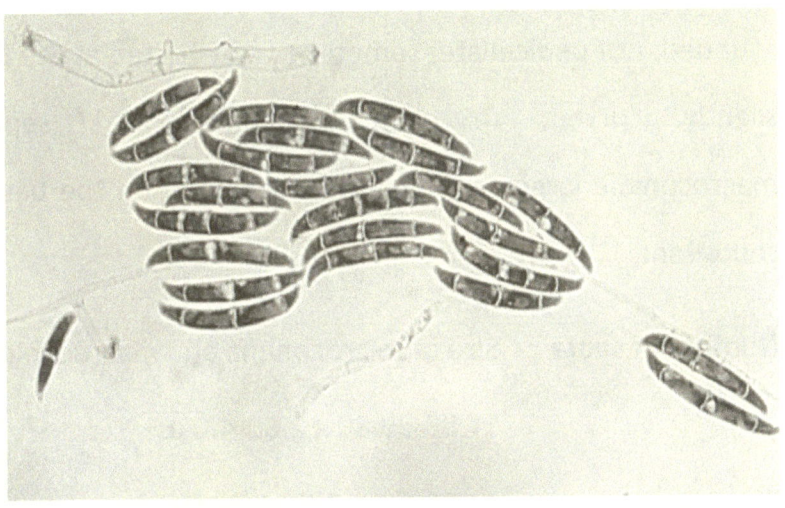

Fig. 33. *Fusarium solani*

34. *Fusarium udum* (Berkeley.) Wollenweber Syn. *Fusarium merismoides*, Corda.

Colonies growing moderate on Potato Dextrose Agar medium; mycelium sparsh; arachnoid; hyaline to pink; later forming stroma; turf like; gelatinous; smooth or wrinkled; fasicled, club like or radiating; conidia scattered dusty; pale then rosy; reddish to orange red; later paler, cylindric-spindle shaped; both ends tempin-like to ellipsoidally narrowed or rounded; not pedicellate; sometimes constricted at the base; slightly curved; microconidia smaller, 0-1 septate; macroconidia larger; 3-7 septate; measured on the basis of septation:

Number of septa	Size of microconidia and macroconidia
1	14.50-29.80 x 2.60-4.0 μ
2	24.90-6.0 x 2.50-4.80 μ
3	34.25-60.94 x 3.10-5.0 μ

Chlamydospores globose; terminal or intercalary; single; paired; rarely in chains; measuring 6.50-8.05 μ

diameter or one or two celled 11.80-13.20 x 5.90-6.80 μ in size.

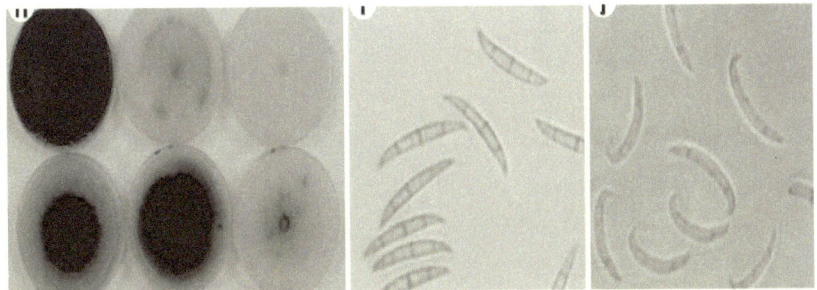

Fig.34 Fusarium udum

35. *Macrophomina phaseolina* (Tassi.) Goid

Colonies fast growing on Potato Dextrose Agar medium; cottony white in young stage; later changing into light grey mycelium; mycelium long; septate and branched; sclerotia common; black; smooth; hard; 100.0 μ to 1.0 mm. in diameter; pycnidia dark brown; solitary or gregarious, 100.0-200 μ in diameter; opening by apical ostioles. pycnidial wall multicellular; with heavily pigmented thick walled cells on the outer side; conidiophores (phialides) hyaline; short; pyriform to cylindrical; 4.90 x 12.70 x 3.49-6.45 μ size: conidia hyaline; ellipsoid to ovoid; 16.80-30.10 x 5.40-9.30 μ in size.

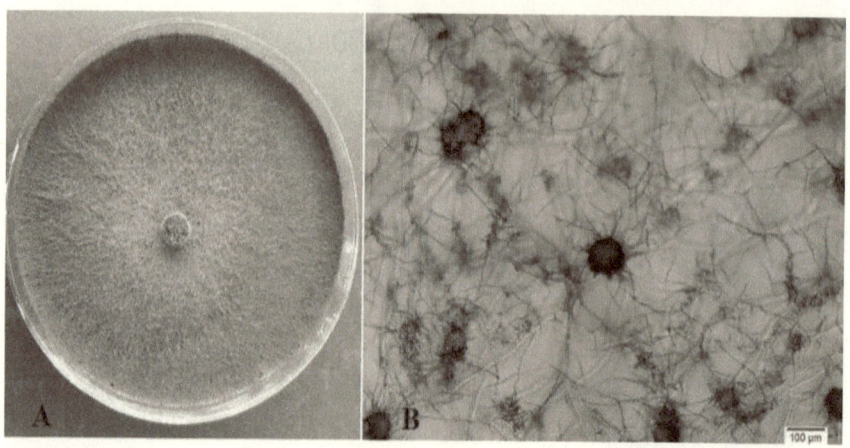

Fig. 35. *Macrophomina phaseolina*

36. *Memnoniella echinata* (Rivolta.) Galloway

Colonies rapidly growing on Potato Dextrose Agar medium; mycelium hyaline; sparse; Sterile hyphae hyaline; septate; conidiophores without foot cell; simple: erect; septate with dark pigment and warted dark granules; measuring 52.48-98.50 x 3.10-4.0 µ in size; usually about seven, conidia one celled; spherical with black pigment; warted; measuring 4.30 µ in diameter; borned in basipetal succession in branched chains.

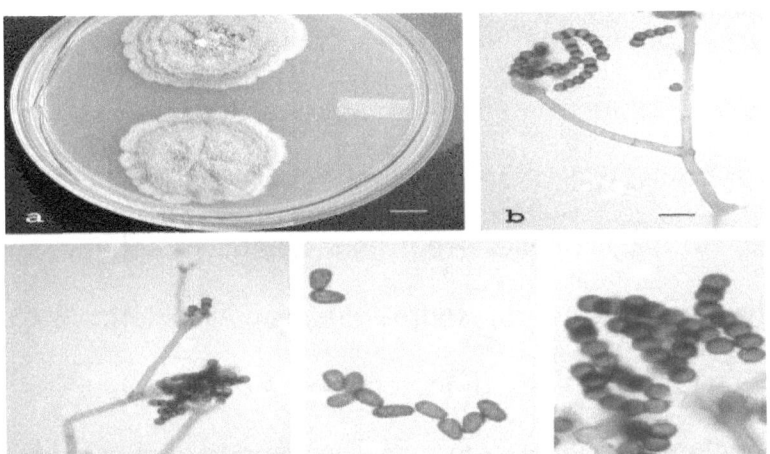

Fig.36. *Memnoniella echinata*

37. *Mucor echinulatus,* Paine Syn. *Mucor circinelloides* Van Tieghem

Colonies rapidly creeping on Potato Dextrose Agar medium; white; cotton in earlier stages; later turn to grayish yellow; reverse colour buff-yellow sporangiophores erect forming a turf; close and deep brown; varying up to 1.0 cm tall; more or less branched in sympodia with branches alternating right and left; short; more or less curved: terminated by a sporangium; secondary braches varying in length or sometimes short giving a sessile appearance of sporangium; sporangia globose; 55.0-80.0 μ in diameter; grey-brown; when walled; erect or slightly incurved; the larger have a diffluent membrane; upper wall persistent in smaller, caduceus; wall of sporangium incrusted persistent firm and smooth; collumellae free; hemispheric or spheric; or oval; colourless; smooth; spores globose or elliptic; 3.10 μ in diameter 4.0-4.80 μ in length; smooth; colourless when single; but pale grey in mass; zygospores globose, exine red brown; covered with very prominent warts; longitudinally striate; chlamydospores smooth: colourless; deep on the length of filament.

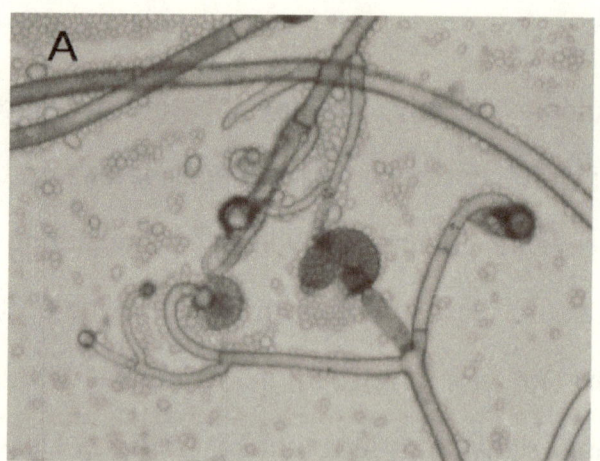

Fig.37. *Mucor circinelloides*

38. *Mucor hiemalis***, Wehmer.**

Colonies fast growing on Potato Dextrose Agar medium; white; cottony in early stage; later changing to grayish yellow; reverse colour buff-yellow. sporangiophores erect; unbranched; sporangia spherical; grey or brownish-yellow with age; measuring 22.35-77.49 µ or 26.40-31.80 to 18.20-29.40 µ; spores unequal; mostly elongated; kidney shaped; or ellipsoid; measuring 3.20-7.50 x 2.50-5.40 µ smooth hyaline with thin membrane, chlamydospores measuring 10.75-25.63 µ in size; zygospores absent.

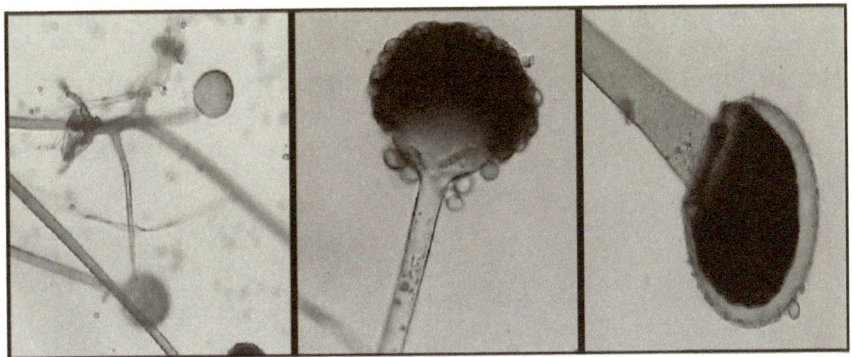

Fig.38. *Mucor hiemalis*

39. *Mucor varians***, Povah.**

Colonies fast growing on Potato Dextrose Agar medium; milky white: whitish yellow to olive buff; reverse colour buff yellow: sporangiophores 9.40- 19.70 μ in diameter: little or profusely branched: much coiled; twisted or interwind forming dense cottony turf with proliferations of hyphae; collumellae often present; sporangia globose or sub-globose; smooth; measuring 64.20-76.80 μ in diameter; yellow or pale orange at first then dark grey at maturity; collumelae free or slightly adnate; variable in shape; sub globose; hemispherical; fiattend; hemispherical; oval; cylindrical; elliptical; conical; small collumellae cylindrical to pyriform; measuring 2670-49.0 x 22.50-44.80 μ in size, membrane tinged grey spores unequal; oval to sub-elliptical measuring 3.90-5.80 x 2.70-4.0 μ in size, zygospores absent.

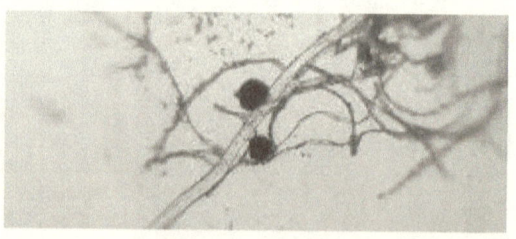

Fig.39. *Mucor varians*

40. *Myrothecium roridum*, Tode.

Colonies fast growing on Potato Dextrose Agar medium; cottony white in young stage; changing to light grey; mycelium long; slender; septate; hyaline; branched: measuring 2.90-4.27 µ with an average of 3.50 in thickness; sporodochia sessile; flattened or irregular; measuring 0.1 mm. x 5.0 mm in diameter; often becoming confluent in larger masses; white green coloured in beginning; turning to black; septa absent; sporodochia arise from a pseudoparenchymatous stromata composed of conidiophores; conidiophores erect; septate; unbranched or forked bush like; main axis tapering; 3-4 celled; basal cell of conidiophores measuring 27.60 x 2.30 µ in size while apical cell measuring 7.10-10.40 x 1.0-1.40 µ with an average of 8.45 x 139 µ in size; branched; 1-2 celled; either singly or in pairs or in whorls; terminating in a whorl of phialides; clavate in whorls of 2-8; a closely packed synnema; conidia produced from the whorl of phialides; clustered forming dark black compact mass on the surface of synnema; elliptical; oblong;

cylindric; truncated at both the ends, single celled; smoky; olive green; with two oil drops; measuring 5.68-8.15 x 1.49-2.45 µ with an average of

9.72 x 2.15 in size; held together in gelatinous mass.

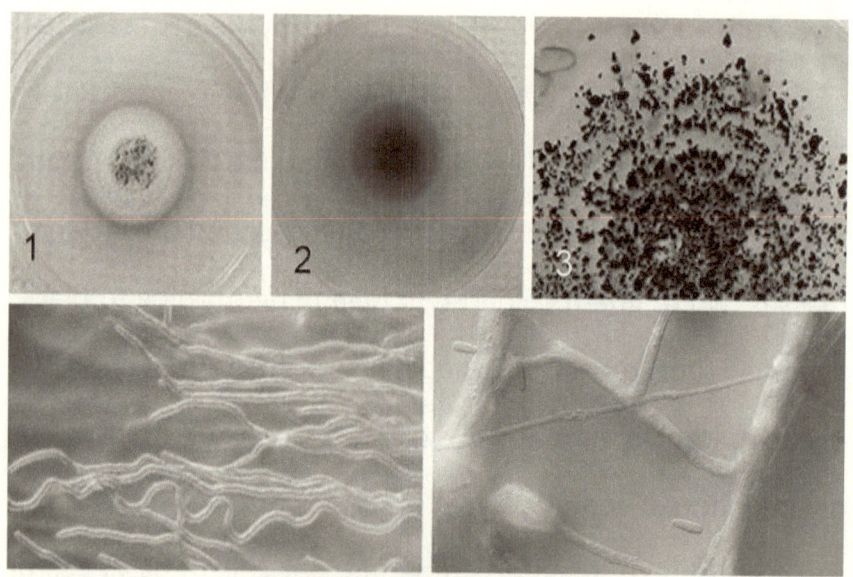

Fig. 40. *Myrothecium roridum*

41. *Nigrospora oryzae*, Hudson

Colonies on Potato Dextrose Agar medium moderately growing; cottony white in beginning; later turning to smoky grey colour or black on abundant sporulation; mycelium hyaline; branched; septate, on maturity brown; 2.70-3.45 µ in width; conidiophores short; ampulliform; hyaline; bearing a single conidium at the tip, 5.36- 7.20 µ in length and 4.30-6.10 µ in width; conidia simple smooth; spherical or ellipsoidal; black or very dark; brown; opaque shining; measuring; 11.50-11.95 µ diameter; o- septate; compressed dorsiventrally.

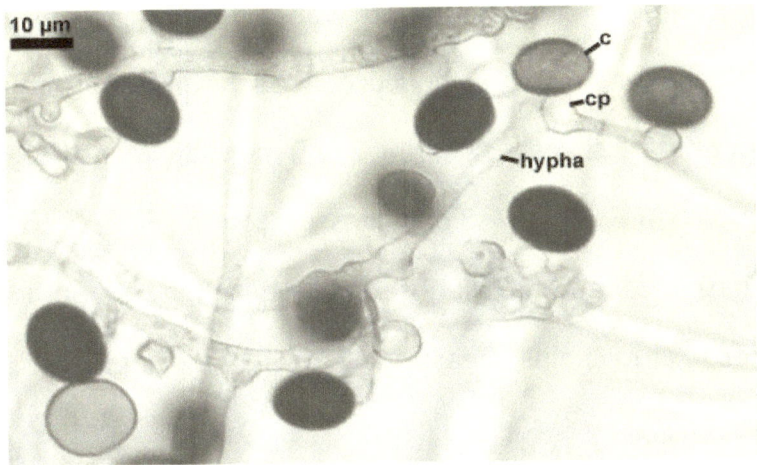

Fig. 41. *Nigrospora oryzae*

42. *Penicillium corylophilum,* Dierck Syn. *P. umbonatum* Shopp.

Colonies on Czapek's agar medium growing moderately with closely woven basal felt; tyough; become velvety; strongly furrowed radially; in earlier stages, white to cream; later become bluish green through its various shades; green or glaucous grey to olive brown; zonate; exudates small; colourless droplets; reverse light brown to fuscous; conidiophores arise as aerial branches from interwoven hyphae; unbranched; smooth walled measuring 52.60-96.20 x 2.10-2.30 µ in size; conidia produced in three stages; typically of primary branches bearing 2 or 3 metulae; measuring 11.95-2.35 µ in size with 4-8 phialides; measuring 6.70-11.52 x 1.90-2.10 µ; conidia sub-globose to elliptical; long; smooth walled; loose or in chains, measuring 2.50-2.90 µ in length.

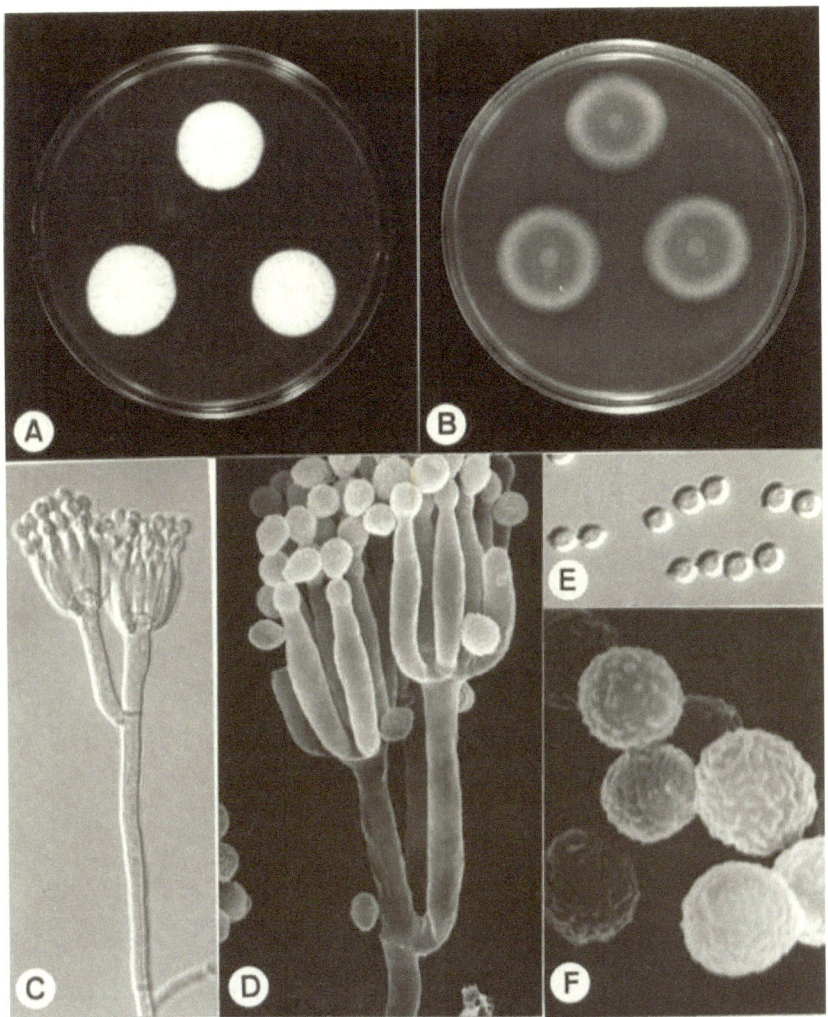

Fig.42. *Penicillium corylophilum*

43. *Penicillium chrysogenum*, Thom. Syn. *P. baculatum*, Westling, *P. roseociterum*, Biourge.

Colonies on Czapek's Agar medium growing; grey green or mixed green; exudates yellow coloured; changing brown to dark brown in colour in old cultures; cottony subfloccose; spreading with a Sterile margin in earlier stages; reverse colour yellow; medium uncoloured; conidiophores arise on aerial branches from the substratum; separate; long or short branches; up to 100.0-200.0 μ long with 1 or 2 alternate divergent branches; phialides 6.90-8.10 x 2.0-2.36 μ conidia pale green to olive green or almost green; elliptical to globose; measuring 3.50-4.30 μ in size.

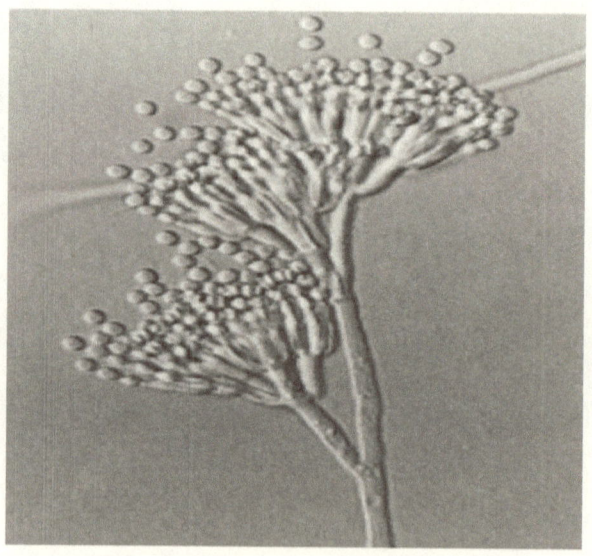

Fig.43. *Penicillium chrysogenum*

44. *Penicillium citrinum*, Thom Syn. *P. aurifluum*. Biourge.

Colonies on Czapek's Agar medium bluish-green to

clear green; changing olive to brownish oliven when old; with Sterile white margin; reverse yellow; conidiophores arise as aerial part of colony in dense except in centre where tufts of aerial hyphae arise: conidiophores arise separately from submerged hyphae or from mycelium on the surface; measuring up to 150.0 μ in length; in two stages; metulae 18.50-29.40 x 1.50-3.0 μ in size; enlarged at the apex up to 5.0 μ; producing a vertical of phialides measuring 6.10-6.90 x 2.20-2.70 μ in size; conidial chains in columns; a septate column arising from each vertical of cells; causing fructification to appear double; triple or more complex; conidia globose; 2.50-2.94 or 3.50 μ; green slightly green.

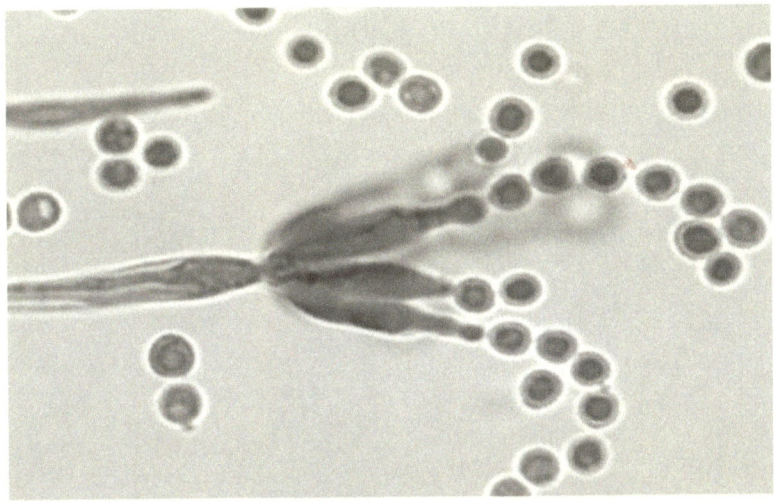

Fig.44. *Penicillium citrinum*

45. *Penicillium expansum* (Link.) Thom.

Colonies on Czapek's agar medium slow in growth; green or grey green; growing vigorously; changing to brown in old cultures; floccose; glaucous; with well defined concentric zones; reverse brown colour in medium; conidiophores emerge directly from the substratum as aerial hyphae or sometimes grouped to form coremia of aerial hyphae or lateral branches; short or long; up to 1 .0 mm. conidia produced in three stages; measuring 128.0-200.30 x 54.80-56.40 µ in size; consisting of primary branches bearing metulae; swollened at the tip; phialides measuring 6.75 x 2.90-3.20 µ in size; conidia yellowish-green to green; elliptical to globose; 2.60-3.20 or 2.80-3.40 µ in size.

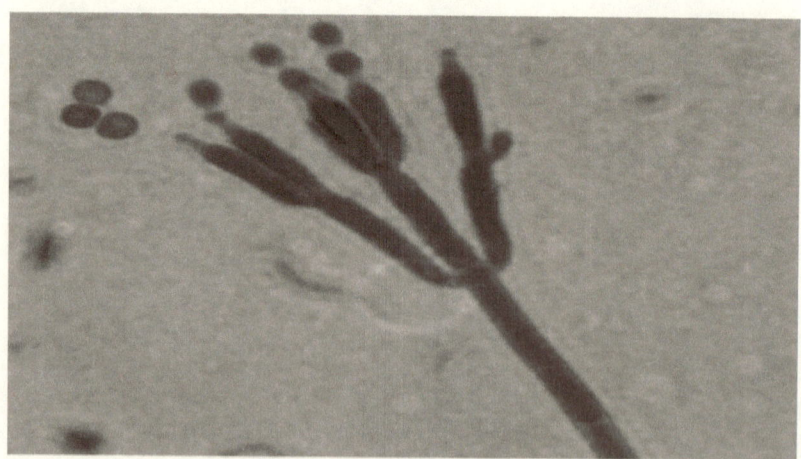

Fig.45. *Penicillium expansum*

46. *Penicillium oxalicum*, Thom.

Colonies on Czapek's Agar medium slow to moderate rate of growth: in shades of blue green; dusky green to dark yellowish green; reverse of the colony green to colonial buff: uncoloured or only slightly coloured; mycelium usually not plentiful but sporulation freely giving a colony; usually in some shades of green; spores produced in a broom like penicillus; typically biverticellate and asymmetric; light yellowish green; measuring 15.80-21.0 x 4.90-11.3 µ in size; borne on smooth walled conidiophores; arising directly from the substratum; conidiophores almost hyaline; sparsely septate; up to 200.00 x 3.5-5.30 µ in size; bearing metulae; often in groups of termninal clusters of 5-6; measuring 88.50-160.50 x 3.0-5.40 µ; conidial heads columner, usually up to 213.0 µ long; metulae usually in groups or in terminal clusters of 5-6 measuring 7.60-12.40 x 2.90-3.50 µ; conidia at first cylindrical; then elliptical; smooth; light yellowish green or pale green in mass; 3.20-5.40 x 2.30-3.50 in size; cleistothecia not observed.

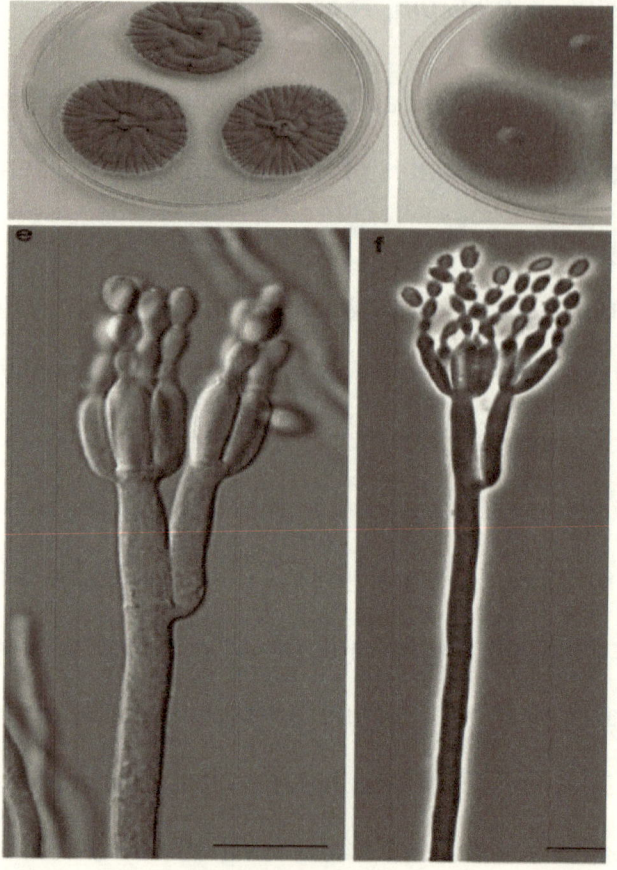

Fig.46. *Penicillium oxalicum*

47. *Penicillium rubrum*, Stoll

Colonies on Czapek's Agar medium growing vigorously; floccose; green to pale green with tentle green margin with green conidia with yellow or olive yellow coloured mycelium; reverse yellowish to red in old cultures exudates in four drops present; conidiophores emerge directly from the substratum or as very shor lateral branches of aerial hyphae; measuring 15.42-29.30 x 3.03-3.40 μ in size; slightly swollened at the apex; biverticillate and symmetrical: measuring 6.90-21.10 μ in size; slightly swollened at the apex. measuring 6.30-12.47 x 1.25-2.30 μ size: conidia smooth; elliptical to globose; measuring 3.20-5.30 x 2.55-3.20 μ; yellowish green to smooth.

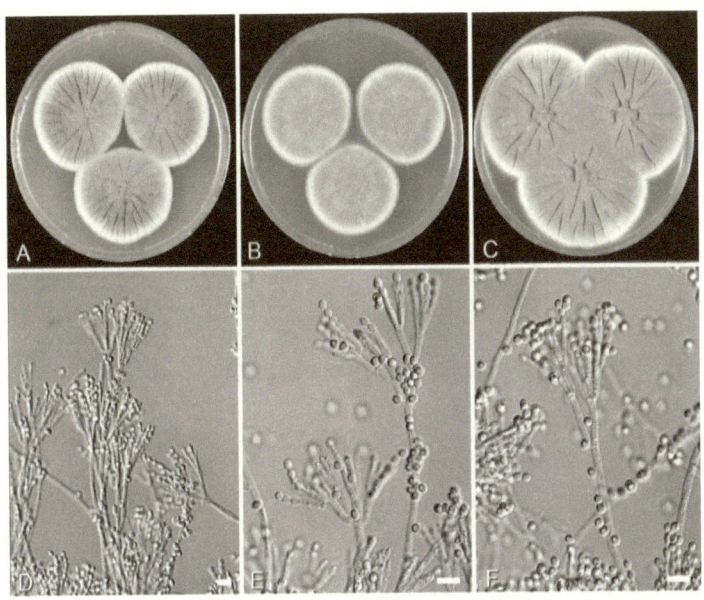

Fig.47. *Penicillium rubrum*

48.*Phoma humicola*, Gilman and Abbott

Colonies fast growing on Potato Dextrose Agar

medium: dark brown spreading; submereged with aerial hyphae; pycnidia dark brown to black or almost carbonaceous black; ostiolate; lenticular or globose, slightly lens shaped with a small papilla at the apex; membranous or Leathery; sub-globose to pyriform with a short neck; measuring 96.50-147.80 x 119.30-124.70 μ in size; conidiophores thread like; short but rare; simple or forked: conidia oblong or bacillate with rounded ends; hyaline; measuring 8.64-11.35 x 3.15-3.85 μ in size.

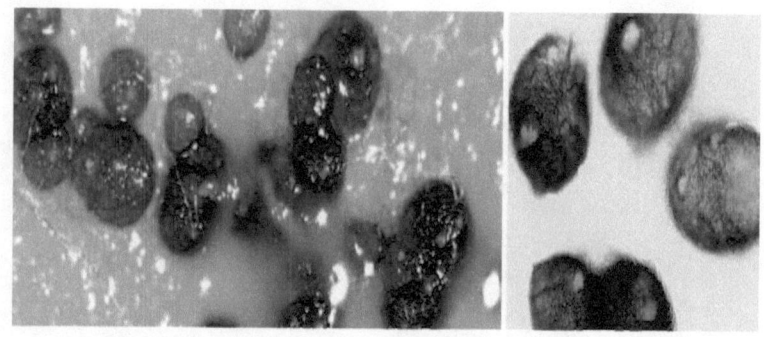

Fig.48.*Phoma humicola*

49. *Rhizoctonia bataticola* (Taub.) Butl.

Colonies on Potato Dextrose Agar medium fast growing; compact: fluffy: abundant aerial mycelium of milky white in earlier stage, which later turn into dark grey to olivaceous brown containing dark brown to black sclerotia; reverse of the colony dark brown; vegetative hyphae septate; hyaline when young; later changed into dark brown or black in colour, 1.75-6.80 μ in thickness; branched; branches at right angle with a constriction at the point of branching along with a septum; sclerotia abundant; variable in shape and size; spherical to globose; olivaceous brown to almost black in colour; 80.50-221.90 μ in diameter either aggregated in masses or formed singly; chlamydospores present in chains; intercalary; spherical: 10.50-22.30 μ oval to dome shaped; 10.80-21.90 x 7.40-14.20 μ thin walled; hyaline or brown in colour; sporodochia absent.

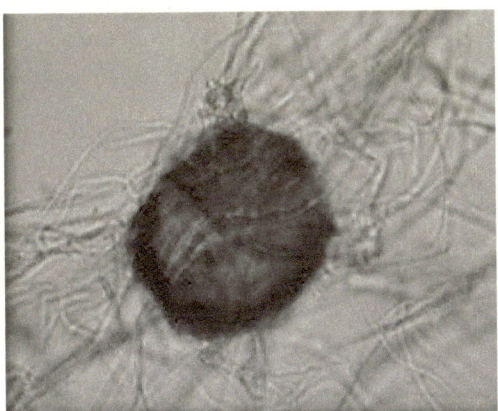

Fig.49. *Rhizoctonia bataticola*

50. *Rhizoctonia solani*, Kuhn

Colonies on Potato Dextrose Agar medium fast growing; ltyaliae: aeriai mycelium; flaky white to buffy; when dry semitranslacent; later pinkish buff to cinnamon brown; with abundant sclerotia; reverse colour of the colony pale ochracepus buff; mycelium septate; branching at the right angles with a constriction at the point of branching or the first septum placed a few microns beyond the point of origin; hyphal branches anastomose; main branch hyphae 3.25-6.90 μ in diameter; short side branch hyphae 4.70-7.20 μ in thickness; sclerotia submerged: clearly visible from the underside of dish: first hyaline; later cream coloured; varying

in shape; round to irregular; hard: frequently aggregated in groups: 0.63 mm. - 2.50 mm in diameter; asexual fruiting bodies and spores absent chlarnydospores hyaline; brownish; in groups; variable in shape and size in early stages; cylindrical or barrel shaped; produced in chains; measuring 17.20-29.50 x 9.85-14.50 μ in size; sporodochia not observed.

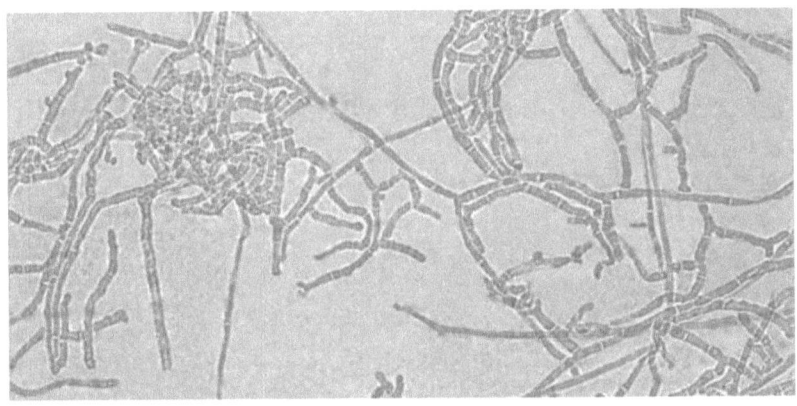

Fig.50. *Rhizoctonia solani*

51. *Rhizopus arrhizus,* Fischer

Colonies on Potato Dextrose Agar medium fast growing; stolons creeping; little developed; nodes not formed regularly; recurring to the substrate in the form of arachnoid hyphae which are strongly raised and distant from the substrate; less exuberance; felt is clearer and not extent into the substrate; rhizoi-ds pale; develop at the nodes and bear sporangia or sometimes in determinately; sporangiophores prostrate; single; forming umbels or corymbs or in their stolons; measuring 0.90-1.80 mm. in length bearing sporangia; larger or smaller; spherical; 126.0-247.80 μ in diameter; collumellae spherical; flattened on the apophysis, 43.40-74.50 μ in length and 65.10-98.30 μ in width; membrane brown; smooth; spores round or oval or presenting obtuse angles; greyish brown; walls striated longitudinally; 4.90-6.80 x 4.60-5.40 μ in size.

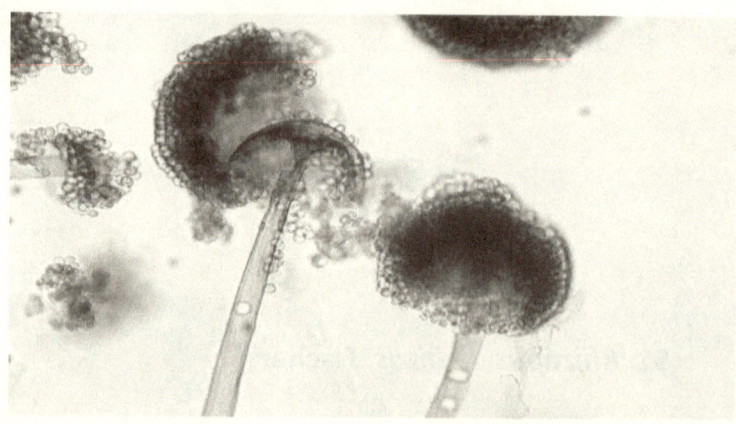

Fig. 51. *Rhizopus arrhizus*

52. *Rhizopus nigricans*, Ehrenberg

Colonies fast growing on Potato Dextrose Agar medium; stolons creeping; recurving to substrate in the form of arachnoid hyphae which are strongly raised and distant from the substrate; implanted at each node by means of rhizoids; internodes often attain a length of 1.0-3.0 cm; hyphae branched; sporangiophores rarely single; united in groups of 3-5, 0.6-3.80 mm in height 26.80-41.30 μ in diameter; apophyses broad; sporangia hemispherical; 100.0-340.0 μ in size; collumellae broad; hemispherical; depressed; 68 μ in diameter x 85.30 μ in height; spores unequal; irregular; round or oval; angular, striate, 9.50-11.80 x 7.60-8,0 μ in size; grey blue; zygospores round or oval, 180.20-218.90 in diameter; exine brown black; verrucose; suspensors swollen, usually unequal; azygospores present; chlamydopores not observed.

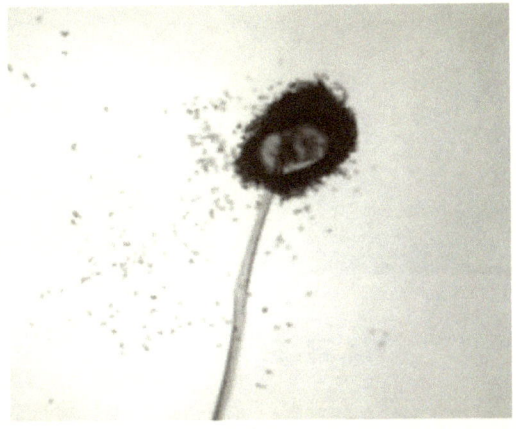

Fig.52. *Rhizopus nigricans*

53. *Sclerotium rolfsii* Saccardo.

Colonies on Potato Destrose Agar medium fast growing; mycelium densely floccose; not ropy; hyphae septate; hyaline or white in young stages but become olive brown to dark brown or olive brown; sclerotia globose; elongated; swollen or flattened; often band like; single or confluent; sometimes covering wide surfaces; pinkish buff to olive brown; hard when dry; internally bright coloured 0.73-2.60 mm. in diameter; rind tissue markedly differentiated from the interior by colour and cell structure; asexual fruiting bodies and spores absent; sporodochia absent.

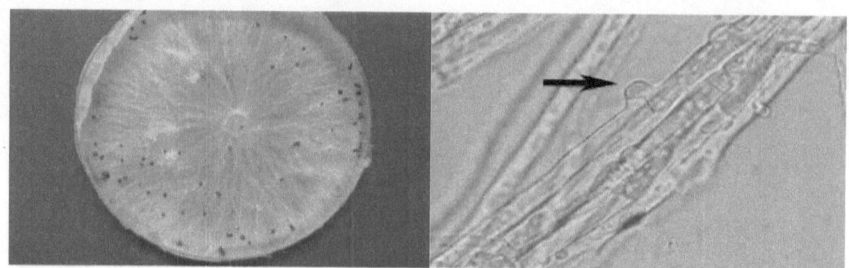

Fig.53. *Sclerotium rolfsii*

54. *Sclerotinia sclerotiorum*.

Colonies on Czapek's Agar medium fast growing;

floccose; vegetative hyphae septate; hyaline; branched; white when young; measuring 8.70-13.50 μ in width; no true conidia were observed; numerous microconidia produced on short chains on stalks; stalks short; 4.90-14.60 x 1.50-3.40 μ in size; due to age in old cultures when food was completely exhausted sclerotia produced; smaller usually round; larger ones round; later long shaped or irregular in shape or 10.0 mm. but, usually 2.90-60.0 mm. in diameter; walls dark brown to black in shade.

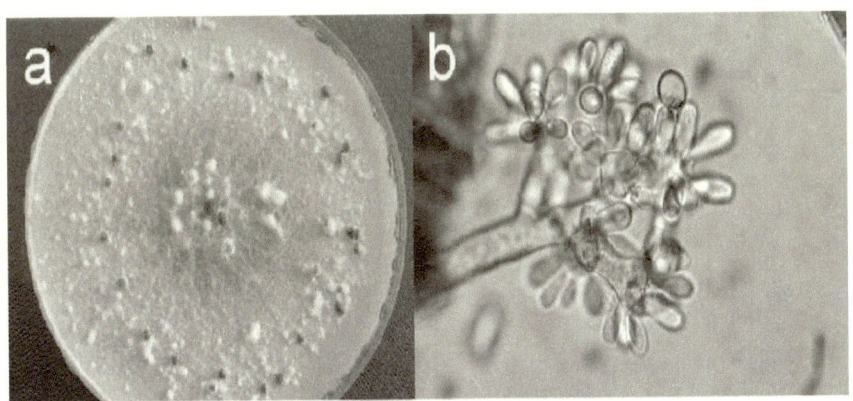

Fig.54. *Sclerotinia sclerotiorum*

55. *Stachybotrys atra*, **Corda.**

Colonies spreading on Potato Dextrose Agar medium: at first hyaline becoming black with age; mycelium creeping; spreading over the substratum: branched; hyaline; septate 5.10-6.0 p thick; with branches almost at right angles: chlamydospores intercalary; oval; ellipsoidal or g1obose; up to 12.0 μ in diameter; articulate with age; conidiophores arise as aerial branches of the mycelium; erect; fuliginous near the apex; hyaline near the base; branched; septate; measuring 67.80- 73.50 x 2.30-3.80 μ in size; slightly alternate towards the apex; bearing at the apex of main stalk a whorl of papillate phialides; phialides measuring 10.2-11.90 x 4.60-5.0 μ in size; non-septate; hyaline or slightly dark coloured; singly or grouped: conidia borned singly on the points of phialides; smooth: elliptical; with acute ends and two oil drops; slightly coloured; when young fuliginous and almost black when mature.

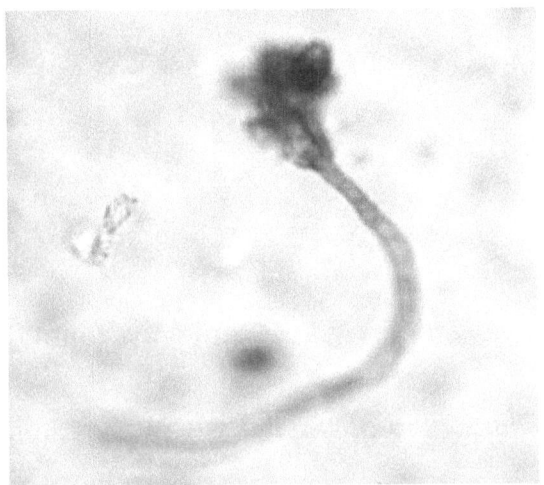

Fig.55. *Stachybotrys atra*

56. Sterile mycelium

Colonies on Potato Dextrose Agar medium moderately growing; fluffy; white; smooth; cream yellow in colour; mycelium branched; septate; hyaline but septa at very long distances; slender and thin 11.20-15.0 μ wide.

A   B

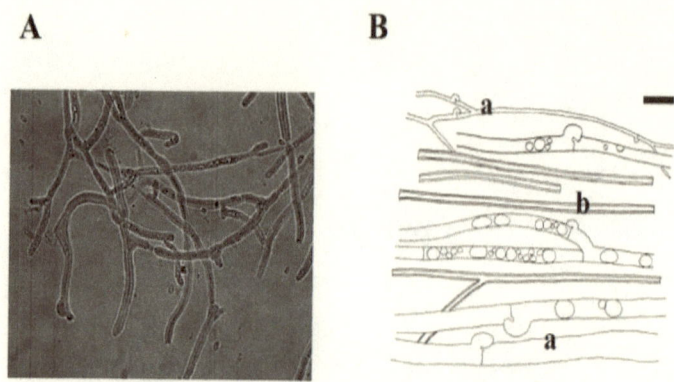

Fig.56. Sterile mycelium

57. *Verticillium albo-atrum* Reinke and Berthold

Colonies fast growing on Czapek's agar medium; vegetative hyphae spreading; hyaline; light coloured or brownish; conidiophores emerge out from the substratum; aerial; erect; simple; dark coloured; paler at the apex; with upto eight whorls; 3-5 branches in the whorl; branches septate; simple or further branched in whorls; terminal branchlets thickened at the base and narrowed at the apex; erect; conidia borned singly on the branchlets: elongated; hyaline or brown coloured egg- shaped; measuring 5.80-12.30 x 2.90-310 in size.

Colonies on Czapek's Agar medium fast growing;

cottony; vegetative hyphae creeping; septate; hyaline or light coloured or brownish dark; brown resting mycelium formed only in association; microsclerotia arise centrally in culture plates; dark brown to black; orulose or botryoidal consisting of swollen globular cells; multiplying by budding; variable in shape; enlongated to irregularly spherical; measuring 17.0-65.0 (100.0) μ in size; conidia are singly at the apex of phialides; elliptical; measuring 2.40-7.80 x 1.50-3.45 μ in size; chlamydospores not observed.

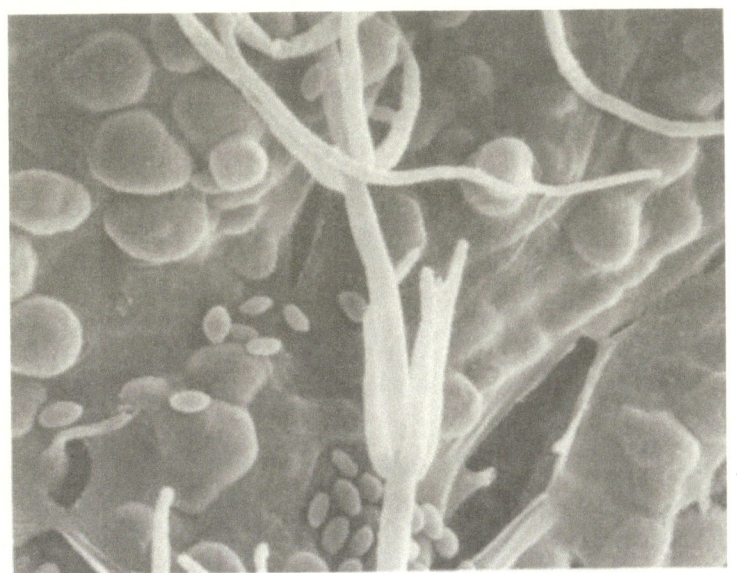

Fig.57. *Verticillium albo-atrum*

****